ALLE ZEIT WACH 1842

Vijay Kumar Stokes

Theories of Fluids with Microstructure

An Introduction

With 44 Figures

Springer-Verlag
Berlin Heidelberg New York Tokyo 1984

Dr. Vijay Kumar Stokes

Department of Mechanical Engineering
Indian Institute of Technology Kanpur, India
and
Corporate Research and Development
General Electric Company
Schenectady, New York, USA

ISBN-13:978-3-642-82353-4 e-ISBN-13:978-3-642-82351-0
DOI: 10.1007/978-3-642-82351-0

Library of Congress Cataloging in Publication Data.
Stokes, Vijay Kumar, 1939 -. Theories of Fluids with microstructure.
Bibliography: p. Includes index. 1. Fluid mechanics. 2. Fluids. I. Title.
TA357.S76 1984 620.1'06 84-14018.

Softcover reprint of the hardcover 1st edition 1984

2061/3020-543210

पूज्यचरणेभ्यः श्रीमत्पितृभ्यः
सादरं समर्पितः

To my parents

Preface

This book provides an introduction to theories of fluids with microstructure, a subject that is still evolving, and information on which is mainly available in technical journals. Several approaches to such theories, employing different levels of mathematics, are now available. This book presents the subject in a connected manner, using a common notation and a uniform level of mathematics. The only prerequisite for understanding this material is an exposure to fluid mechanics using Cartesian tensors.

This introductory book developed from a course of semester-length lectures that were first given in the Department of Chemical Engineering at the University of Delaware and subsequently were given in the Department of Mechanical Engineering at the Indian Institute of Technology, Kanpur.

The encouragement of Professor A. B. Metzner and the warm hospitality of the Department of Chemical Engineering, University of Delaware, where the first set of notes for this book were prepared (1970-71), are acknowledged with deep appreciation. Two friends and colleagues, Dr. Raminder Singh and Dr. Thomas F. Balsa, made helpful suggestions for the improvement of this manuscript.

The financial support provided by the Education Development Centre of the Indian Institute of Technology, Kanpur, for the preparation of the manuscript is gratefully acknowledged.

Special mention must be made of Julia A. Kinloch, who typed the manuscript and whose patience and professionalism made the author's task much easier. The help of David D. Raycroft, who gave valuable editorial advice, and Maria A. Barnum, who prepared the galleys, is also gratefully acknowledged.

Finally, this work would not have been possible without the patience shown over many years by Prabha, Chitra, and Anuradha.

Schenectady, New York, 1984 VIJAY K. STOKES

Contents

CHAPTER 4
Anisotropic Fluids

CHAPTER 5
Micro Fluids

CHAPTER 6
Micropolar Fluids

APPENDIX

Introduction

This book provides an introduction to the theories of fluids with microstructure. Flows of such fluids can exhibit many effects that are not possible in classical nonpolar Stokesian fluids. This material attempts to present a connected account of three different types of theories in a common notation. In keeping with the introductory nature of the presentation, only Cartesian tensors have been used. The level of presentation only assumes an exposure to fluid mechanics using Cartesian tensors. The notation is explained in the Appendix, where some of the results, which are required, have also been summarized.

In addition to the usual concepts of nonpolar fluid mechanics, there are two main physical concepts that go into building theories of fluids with microstructure: couple stresses and the concept of internal spin. Couple stresses are a consequence of assuming that the mechanical action of one part of a body on another, across a surface, is equivalent to a force and a moment distribution. In classical nonpolar mechanics, moment distributions are not considered, and the mechanical action is assumed to be equivalent to a force distribution only. The laws of motion can then be used for defining the stress tensor which, necessarily, turns out to be symmetric. Thus, in nonpolar mechanics, the state of stress at a point is defined by a symmetric second order tensor which is a point function that has six independent components. However, in polar mechanics the mechanical action is assumed to be equivalent to both a force and a moment distribution. The state of stress is then measured by a stress tensor and a couple stress tensor. In general, neither of these second order tensors is symmetric, so that the state of stress at a point is measured by eighteen independent components. Thus, the concept of couple stresses results from a study of the mechanical interactions taking place across a surface and, conceptually, is not related to the kinematics of motion.

On the other hand, the concept of microstructure is a kinematic one. For classical fluids without microstructure, all the kinematic parameters are assumed to be determined, once the velocity field is specified. Thus, if the velocity field is identically zero, then there is no motion, and the linear and angular momenta of all material elements must also be identically zero. However, even when the velocity field is zero, the angular momentum may be visualized as being nonzero by "magnifying" the continuum picture until the identities of "individual" particles can be "seen." A particle may not have any velocity of translation, so that its macroscopic velocity and thus its linear momentum is zero, but the particle may be spinning about an axis.

This spin would give rise to an angular momentum. If the same phenomenon is assumed to hold true at the continuum level, then angular momentum can exist even in the absence of linear momentum. This is not true in the classical theories of fluid mechanics. At the kinematic level, a specification of the velocity field is then not sufficient, and additional kinematic measures, independent of the velocity field, must be introduced to describe this internal spin. Such a fluid is said to have microstructure.

The concepts of couple stresses and microstructure are conceptually different. The first concept has its origins in the way mechanical interactions are modeled, while the second one is essentially a kinematic one, and arises out of an attempt to describe point particles having "structure." Whereas in a general theory of fluids with microstructure, couple stresses and internal spin may be present simultaneously, theories of fluids in which couple stresses are present, but microstructure is absent, are also possible. Similarly, microstructure may be considered in the absence of couple stresses. In this way the main consequences of each of these concepts may be studied before proceeding to the study of more general theories.

Several different approaches may be used for formulating such theories. For example, a statistical mechanics model, which assumes noncentral forces between particles, is known to give rise to couple stresses. Thus a continuum theory for fluids with microstructure may be obtained from such a model. The concept of couple stresses may also be introduced purely on the basis of a continuum argument. Microstructure can be introduced heuristically. Such theories may also be formulated by an averaging procedure in which the macroscopic variables are obtained by taking suitable averages over continuum domains. In this book an attempt has been made at unifying three different ways of developing such theories.

The simplest theory of couple stresses in fluids, in the absence of microstructure, is given in Chapter 3. The most important effect of couple stresses is to introduce a size-dependent effect that is not predicted by the classical nonpolar theories. For example, in pipe Poiseuille flow, even after all the variables have been nondimensionalized in the usual way, the velocity profile is a function of the pipe radius.

Chapter 4 describes a simple theory of fluids with microstructure in which couple stresses are absent. The inclusion of microstructure predicts phenomena such as stress relaxation and Bingham-plastic-like flow.

Finally, more general theories, in which both couple stresses and microstructure are accounted for in a systematic manner, are discussed in Chapters 5 and 6.

Several available approaches to formulating theories of fluids with microstructure have not been discussed. Neither has an attempt been made to compare all the theories that are now available. References to several excellent review articles, which compare different theories, are given below. In particular, the article by S. C. Cowin provides a detailed self-contained presentation of his theory of Polar fluids. The last two references give a fairly complete account of the polar fluid theories proposed by A. C. Eringen.

Pertinent references have been listed at the end of each chapter. Finally, a comprehensive bibliography, in chronological order, is given at the end of the book.

References

1. Ericksen, J. L. (1967). Continuum Theory of Liquid Crystals, *Appl. Mech. Rev.* **20**, 1029-1032.
2. Ariman, T., Turk, M. A., and Sylvester, N. D. (1973). Microcontinuum Fluid Mechanics — A Review, *Int. J. of Eng. Sci.* **11**, 905-930.
3. Ariman, T., Turk, M. A., and Sylvester, N. D. (1974). Review Article: Applications of Microcontinuum Fluid Mechanics, *Int. J. Eng. Sci.* **12**, 273-293.
4. Cowin, S. C. (1974). The Theory of Polar Fluids, "Advances in Applied Mechanics" (Chia-Shun Yih, Ed.), Vol. 14, Academic Press, New York, 279-347.
5. Eringen, A. C. (Editor), (1976). "Continuum Physics, Vol. IV: Polar and Non-Local Field Theories," Academic Press, New York.
6. Eringen, A. C. (1980). Theory of Anisotropic Micropolar Fluids, *Int. J. Eng. Sci.* **18**, 5-17.

CHAPTER 1

Kinematics of Flow

1.1 Introduction

This chapter examines some aspects of kinematic measures of motion. The discussion is limited to the classical model, in which microstructure is not considered, and where the motion is specified by the history of the particle velocity, so that there is no motion if the velocity field is identically zero. Theories in which there is motion even in the absence of a velocity distribution are considered in later chapters.

In the usual discussion on the kinematics of flow, an emphasis is placed on the rates of strain of material line elements and on the rates of shear strain between two orthogonal material line elements. The rates of rotation of line elements are also considered. However, the study of kinematics does not usually go beyond a consideration of the rate of deformation tensor, which determines the normal and shear strain rates, but does not measure all aspects of motion. Some higher order kinematic measures are introduced in this chapter.

Cartesian tensors have been used for developing the theory. In some instances a matrix representation for tensors has been used for convenience. A summary of the main results concerning Cartesian tensors, as well as a description of the notation, is given in the Appendix.

1.2 Velocity Gradient Tensor

The state of motion of a continuous body is assumed to be completely specified when the velocity distribution is known. Once the velocity $\mathbf{v}(\mathbf{x},t)$ of the particle at $\mathbf{x}$ is known at time t, a study of the deformations of line elements may be attempted. Since deformations of material elements can be very large over finite time intervals, a study of deformations over infinitesimal time intervals is appropriate. This leads naturally to a study of the rates of deformation rather than deformations.

Consider a mass of fluid that has the configuration B at time t and B' at time $t + \Delta t$. Let $\overrightarrow{PQ} = d\mathbf{x} = d\xi\mathbf{n}$ be a material line element in B which, at time t, is at the point $\mathbf{x}$, where $\mathbf{n}$ is a unit vector along PQ, as shown in Fig. 1.2.1. After time Δt, the same material line element, which will now be in configuration B', will move to $\overrightarrow{P'Q'} = d\mathbf{X} = d\Xi\mathbf{N}$, where the point P' has the position vector $\mathbf{X}$, and $\mathbf{N}$ is a unit vector along $P'Q'$. The general problem of kinematics is then to relate the line element $d\mathbf{X}$, at time $t + \Delta t$, to its original state $d\mathbf{x}$, at time t. Thus, the problem of studying short time deformations is to develop a relation of the form $d\mathbf{X} = \mathbf{f}(d\mathbf{x},\Delta t)$. Let $\mathbf{v}$ and $\mathbf{v} + d\mathbf{v}$ be the velocities of the points P and Q, respectively, so that $\vec{PP'} = \mathbf{v}\,\Delta t$ and $\vec{QQ'} = (\mathbf{v} + d\mathbf{v})\Delta t$. Then, from the geometry shown in Fig. 1.2.1

$$d\mathbf{X} = d\mathbf{x} + d\mathbf{v}\,\Delta t \tag{1.2.1}$$

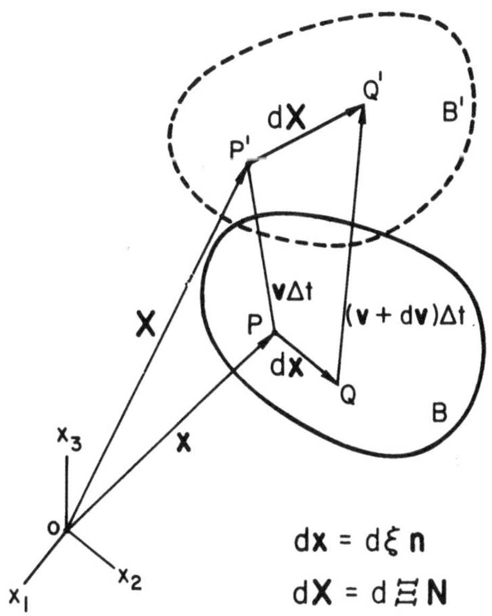

Fig. 1.2.1 Configurations of a body at two times.

Now $d\mathbf{v} = (dv_i)$ is the velocity difference $\mathbf{v}_Q - \mathbf{v}_P$ of the points P and Q at the same time t. Therefore, since at time t the velocity of the different fluid particles is a function of the coordinates x_i of $\mathbf{x}$,

$$dv_i = \frac{\partial v_i}{\partial x_r}\, dx_r = v_{i,r} dx_r$$

Thus

$$dv_i = g_{ri} dx_r \quad or \quad d\mathbf{v} = \mathbf{G}^T d\mathbf{x} \tag{1.2.2}$$

where $\mathbf{G} = (g_{ij}) = (v_{j,i})$ is called the velocity gradient tensor.

Using this result, Eq. (1.2.1) gives

$$dX_i = (\delta_{ri} + \Delta t\, g_{ri})\, dx_r \quad or \quad d\mathbf{X} = (\mathbf{I} + \Delta t\, \mathbf{G}^T)\, d\mathbf{x} \tag{1.2.3}$$

which shows that, for short intervals of time, the configurations of a material line element can be determined once the velocity gradient tensor $g_{ij} = v_{j,i}$ is known.

The next step is to develop expressions for rates of change of lengths and angles between line elements and the rates of rotation of line elements. Other aspects of deformation, such as the rate of twisting and the rate at which curvature is induced in line elements, will also be considered.

Important roles are played by the symmetric and skew-symmetric parts of $\mathbf{G}$. The symmetric part $\mathbf{D}$, with components $d_{ij} = \frac{1}{2}(g_{ij} + g_{ji}) = \frac{1}{2}(v_{j,i} + v_{i,j})$, is called the rate of deformation tensor and the skew-symmetric part $\mathbf{W}$, with components $w_{ij} = \frac{1}{2}(g_{ij} - g_{ji}) = \frac{1}{2}(v_{j,i} - v_{i,j})$, is called the spin tensor.

1.3 Rate of Deformation Tensor

At time t, let $d\xi$ be the length of a material line element $d\mathbf{x}$ at the point $\mathbf{x}$, in the direct $\mathbf{n}$, so that $d\mathbf{x} = d\xi\mathbf{n}$. After time Δt the material line element becomes $d\mathbf{X}$, having a length $d\Xi$ in the direction $\mathbf{N}$ which, in general, is different from $\mathbf{n}$, so that $d\mathbf{X} = d\Xi\mathbf{N}$. Physically, the time rate of change of length of a material line element, per unit length, is of interest since it indicates the rate at which the line element is being stretched. More formally, the rate of normal strain, $\epsilon(\mathbf{x},t;\mathbf{n})$, at the point $\mathbf{x}$ at time t in the direction $\mathbf{n}$, is defined to be the time rate of change of the length of a material line element, per unit length, which is originally along the direction $\mathbf{n}$. Thus, for a material line element $d\mathbf{x} = d\xi\mathbf{n}$ at time t, which goes into $d\mathbf{X} = d\Xi\mathbf{N}$ at time $t + \Delta t$,

$$\epsilon(\mathbf{x},t;\mathbf{n}) = \frac{1}{d\xi}\frac{D}{Dt}(d\xi) = \underset{\Delta t \to 0}{limit}\ \frac{1}{d\xi}\left[\frac{d\Xi - d\xi}{\Delta t}\right] \tag{1.3.1}$$

where D/Dt denotes the material time derivative.

Now $dX_i = (\delta_{ri} + \Delta t\ g_{ri})dx_r = d\xi(\delta_{ri} + \Delta t\ g_{ri})n_r$ implies that $d\Xi^2 = dX_i dX_i = d\xi^2(\delta_{ri} + \Delta t\ g_{ri})(\delta_{si} + \Delta t\ g_{si})n_r n_s$, which can be written as

$$d\Xi = d\xi[1 + \Delta t(g_{rs} + g_{sr})n_r n_s + (\Delta t)^2 g_{ri} g_{si} n_r n_s]^{1/2}$$

or, since Δt is small,

$$d\Xi = d\xi[1 + \frac{\Delta t}{2}(g_{rs} + g_{sr})n_r n_s + 0(\Delta t^2)]$$

Use of the definition of ϵ given in Eq. (1.3.1) then results in

$$\epsilon(\mathbf{x},t;\mathbf{n}) = d_{rs}n_r n_s = \mathbf{n}^T\mathbf{D}\mathbf{n} \tag{1.3.2}$$

which shows that normal strain rates are determined by the rate of deformation tensor.

Let $\mathbf{n}_1 = (l_i)$ and $\mathbf{n}_2 = (m_i)$ be two mutually orthogonal directions. Then, the rate of shear strain, $\gamma(\mathbf{x},t;\mathbf{n}_1,\mathbf{n}_2)$, between these two directions, is defined as the time rate of the decrease of the angle between two material line elements which are originally along $\mathbf{n}_1$ and $\mathbf{n}_2$, respectively. In order to obtain an expression for the shear strain rate, consider two material line elements which, at time t, are given by $d\mathbf{x} = d\xi_1\mathbf{n}_1$, that is by $dx_i = d\xi_1 l_i$, and $d\mathbf{y} = d\xi_2\mathbf{n}_2$, or $dy_i = d\xi_2 m_i$. After time Δt they will become, respectively, $dX_i = d\xi_1(\delta_{ri} + \Delta t\ g_{ri})l_r$ and $dY_i = d\xi_2(\delta_{si} + \Delta t\ g_{si})m_s$. By definition the angle between $d\mathbf{X}$ and $d\mathbf{Y}$ is $(\pi/2 - \gamma\Delta t)$, so that $dX_i dY_i = d\Xi_1 d\Xi_2 \cos(\pi/2 - \gamma\Delta t) = d\Xi_1 d\Xi_2 \sin(\gamma\Delta t)$, where $d\Xi_1$ and $d\Xi_2$ are, respectively, the lengths of $d\mathbf{X}$ and $d\mathbf{Y}$. Thus

$$\begin{aligned} d\Xi_1 d\Xi_2 \sin(\gamma\ \Delta t) &= d\xi_1 d\xi_2(\delta_{ri} + \Delta t\ g_{ri})(\delta_{si} + \Delta t\ g_{si})l_r m_s \\ &= d\xi_1 d\xi_2[(\Delta t)(g_{rs} + g_{sr})l_r m_s + (\Delta t)^2\ g_{ri}g_{si}l_r m_s] \end{aligned}$$

as $l_r m_r = 0$. Since Δt is small, $\sin(\gamma\Delta t) \cong \gamma\Delta t$, so that

$$\gamma = \underset{\Delta t \to 0}{limit} \frac{d\xi_1\ d\xi_2}{d\Xi_1\ d\Xi_2}[2d_{rs}\ l_r\ m_s + (\Delta t)g_{ri}\ g_{si}\ l_r\ m_s]$$

Now $d\Xi_1 = d\xi_1(1 + \Delta t\ d_{rs}l_r l_s)$, so that $d\xi_1/d\Xi_1 = [1 + \Delta t\ d_{rs}l_r l_s]^{-1}$, which, in the limit, gives $d\xi_1/d\Xi_1 = 1$. Finally

$$\gamma(\mathbf{x},t;\mathbf{n}_1,\mathbf{n}_2) = 2d_{rs}l_r m_s = 2\mathbf{n}_1^T\mathbf{D}\mathbf{n}_2 \tag{1.3.3}$$

Thus, the rate of deformation tensor gives all the information concerning normal and shear strain rates. By choosing the directions, along which these rates are being determined, along the coordinate axes, the following interpretation for $\mathbf{D}$ can be shown to be true: the diagonal components of $\mathbf{D}$ give the normal strain rate along the corresponding axis. For example, d_{11} is the rate of normal strain along the x_1 axis, and so on. On the other hand, the off-diagonal components give half the shear strain rate along the corresponding axes. For example, d_{12} is half the shear rate between the x_1 and x_2 axes, that is $d_{12} = \frac{1}{2}\gamma(\mathbf{x},t;\mathbf{i}_1,\mathbf{i}_2)$. Thus, the rate of deformation tensor is determined once the normal strain rates and shear strain rates are known, respectively, along and between any triple of mutually orthogonal directions.

1.4 Analysis of Strain Rates

The analysis of strain rates is concerned with the following three questions:

(a) At each point in a fluid, are there three mutually orthogonal directions such that the rates of shear strain between each pair are zero? Such a triple of mutually orthogonal directions, if they exist, are called the principal directions of strain rate.

(b) Along which directions does the rate of normal strain have extreme values? And

(c) Between which pair of orthogonal directions is the rate of shear strain a maximum?

Since $\mathbf{D}$ is a symmetric tensor, all its proper values are real and at least one set of three mutually orthogonal proper vectors exists. When the three proper values are distinct, there is only one such set of proper vectors. When two of the proper values are equal, there is one proper vector corresponding to the distinct proper value and every vector normal to it is a proper vector corresponding to the equal proper values. Finally, when the three proper values are equal, every direction is a proper vector of $\mathbf{D}$. Every set of three mutually orthogonal proper vectors of $\mathbf{D}$ constitutes a set of principal axes of strain, for, if $\mathbf{n}_1$, $\mathbf{n}_2$ and $\mathbf{n}_3$ are a set of mutually orthogonal proper vectors, then, $\gamma(\mathbf{n}_1,\mathbf{n}_2) = 2\mathbf{n}_1^T\mathbf{D}\mathbf{n}_2 = 2\gamma_2\mathbf{n}_1^T\mathbf{n}_2 = 0$, since $\mathbf{D}\mathbf{n}_2 = \gamma_2\mathbf{n}_2$, and so on. Thus, at least one set of principal directions of strain rate always exists at each point.

The normal strain rate along the direction $\mathbf{n}$ is given by $\epsilon(\mathbf{n}) = \mathbf{n}^T\mathbf{D}\mathbf{n}$. Extreme values of $\epsilon(\mathbf{n})$ are to be found subject to the constraint $\mathbf{n}^T\mathbf{n} = 1$. These values are determined by $\mathbf{D}\mathbf{n} = \lambda\mathbf{n}$, so that the extreme values of the normal strain rate occur along the principal directions of strain rate.

The extreme values of the shear strain rate may be shown to occur between pairs of directions that bisect the principal axes of strain rate. The corresponding local extreme is given by the difference of the corresponding principal strain rates.

If the coordinate axes are chosen to be parallel to the principal axes of strain rate, the resulting coordinate system is called a principal coordinate system. In a principal coordinate system $\mathbf{D}$ has a diagonal representation, the off-diagonal elements being zero.

1.5 Spin Tensor

The spin tensor $w_{ij} = \frac{1}{2}(g_{ij} - g_{ji})$, which is the skew-symmetric part of the velocity gradient tensor, will now be shown to measure the rates of rotation of line elements in a certain average sense. With the skew-symmetric tensor w_{ij} can be associated the pseudovector $\omega_i = \frac{1}{2}e_{ijk}w_{jk}$, which gives the dual relation $w_{ij} = e_{ijk}\omega_k$. A use of $w_{ij} = \frac{1}{2}(v_{j,i} - v_{i,j})$ results in $\omega_i = \frac{1}{2}e_{ijk}v_{k,j}$ or $\boldsymbol{\omega} = \frac{1}{2}\nabla\times\mathbf{v}$. The pseudovector $\boldsymbol{\omega}$ is called the vorticity.

Let $\mathbf{n}_1 = (l_i)$, $\mathbf{n}_2 = (m_i)$ and $\mathbf{n}_3 = (n_i)$ be a set of right-handed orthonormal vectors. Consider a line element $d\mathbf{x} = d\xi_1 \mathbf{n}_1$, along $\mathbf{n}_1$, which is represented by OA in Fig. 1.5.1. After time Δt, the element $\vec{OA} = d\mathbf{x} = d\xi_1 \mathbf{n}_1$ will deform into $\vec{OB} = d\mathbf{X} = (\mathbf{I} + \Delta t\ \mathbf{G}^T) d\mathbf{x} = d\xi_1 (\mathbf{I} + \Delta t\ \mathbf{G}^T)\mathbf{n}_1$, so that the vector $\vec{AB} = \Delta t\ d\xi_1 \mathbf{G}^T \mathbf{n}_1$. If a perpendicular is dropped from B onto a line through A which is parallel to $\mathbf{n}_2$, meeting it at B', then $AB' = (\vec{AB})^T \mathbf{n}_2 = \Delta t\ d\xi_1 \mathbf{n}_1^T \mathbf{G} \mathbf{n}_2$. The rotation of $\vec{OA}$ about $\mathbf{n}_3$ is given by $\Delta\theta_1 = AB'/OA = \Delta t\ \mathbf{n}_1^T \mathbf{G}\mathbf{n}_2$. Similarly, the rotation $\Delta\theta_2$ of an element $d\mathbf{y} = d\xi \mathbf{n}_2$, along $\mathbf{n}_2$, about $\mathbf{n}_3$, is $\Delta\theta_2 = \Delta t\ \mathbf{n}_2^T \mathbf{G}(-\mathbf{n}_1) = -\Delta t\ \mathbf{n}_1^T \mathbf{G}^T \mathbf{n}_2$. Therefore, the average rotation of $d\mathbf{x}$ and $d\mathbf{y}$ about $\mathbf{n}_3$ is $(\Delta\theta_1 + \Delta\theta_2)/2 = \Delta t\ \mathbf{n}_1^T [\frac{1}{2}(\mathbf{G} - \mathbf{G}^T)]\mathbf{n}_2 = \Delta t\ \mathbf{n}_1^T \mathbf{W}\mathbf{n}_2$. Thus, the average rotation rate of two line elements along $\mathbf{n}_1$ and $\mathbf{n}_2$, about $\mathbf{n}_3$, is given by $\mathbf{n}_1^T \mathbf{W} \mathbf{n}_2 = w_{rs} l_r m_s$. This average rate would seem to depend on the choice of the elements being along $\mathbf{n}_1$ and $\mathbf{n}_2$. However, a use of $w_{rs} = e_{rst}\,\omega_t$, results in $w_{rs} l_r m_s = e_{rst} l_r m_s \omega_t = n_t \omega_t$, since $e_{rst} l_r m_s$ is the cross product $\mathbf{n}_3 = (n_i)$ of $\mathbf{n}_1$ and $\mathbf{n}_2$. The expression $n_t \omega_t = \mathbf{n}_3^T \boldsymbol{\omega}$ shows that this average rate of rotation is the same for every pair of orthogonal line elements in the plane of $\mathbf{n}_1$ and $\mathbf{n}_2$. Now, if $\Omega(\mathbf{n})$ is defined to be the average rate of counterclockwise rotation about $\mathbf{n}$, of any two perpendicular material line elements which lie in the plane normal to $\mathbf{n}$, then

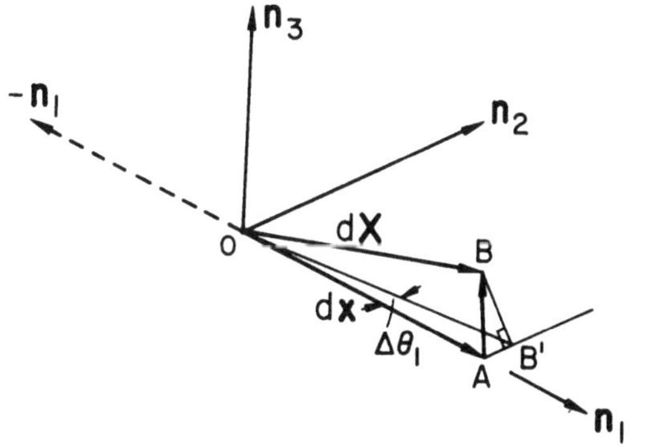

Fig. 1.5.1 Rotation of line elements about the **n** axis.

$$\Omega(\mathbf{n}) = n_r \omega_r = \mathbf{n}^T \boldsymbol{\omega} \tag{1.5.1}$$

Thus $\boldsymbol{\omega}$, and therefore $\mathbf{W}$, measures the average rate of rotation of line elements.

Given $\boldsymbol{\omega}$, or $\mathbf{W}$, the maximum rate of rotation is $\Omega(\mathbf{n}) = |\boldsymbol{\omega}|$ and occurs for orthogonal line elements which lie in the plane perpendicular to $\boldsymbol{\omega}$. On the other hand $\Omega(\mathbf{n}) = 0$ for any $\mathbf{n}$ which is orthogonal to $\boldsymbol{\omega}$.

The components of $\boldsymbol{\omega}$ give the average rates of rotation about the coordinate axes. For example, ω_1 is the average rate of counterclockwise rotation of any two orthogonal material line elements in the $x_2 - x_3$ plane about the x_1 axis.

Section (1.2) showed that $d\mathbf{X} = (\mathbf{I} + \Delta t\, \mathbf{G}^T)\,d\mathbf{x}$, that is, the short term deformations of $d\mathbf{x}$ are determined by the velocity gradient tensor $\mathbf{G}$. Section (1.3) then showed that the rate of deformation tensor $\mathbf{D}$, which is the symmetric part of $\mathbf{G}$, determines the normal and shear strain rates. This section has also shown that the spin tensor $\mathbf{W}$, the skew-symmetric part of $\mathbf{G}$, determines the rates of rotation of line elements in a certain average sense. Thus far the kinematic analysis has followed a pattern which is usual in fluid mechanics. Section (1.6) will show that the measure $\mathbf{D}$ is inadequate for describing certain properties of deformations.

1.6 Curvature-Twist Rate Tensor

Besides undergoing extensions, shearing and rotations that are measured, respectively, by the rate of normal strain, the rate of shear strain, and the vorticity, a line element can undergo twisting and changes in curvature. Measures for these parameters of deformation are developed in this section.

The twist rate $k(\mathbf{x},t;\mathbf{n})$, at a point $\mathbf{x}$, at time t, about the direction $\mathbf{n}$, is defined as the twist rate per unit length of a material line element originally in the direction $\mathbf{n}$. For convenience $k(\mathbf{x},t;\mathbf{n})$ will be abbreviated to $k(\mathbf{n},t)$ and sometimes to just $k(\mathbf{n})$. With reference to Fig. 1.6.1, consider a ma-

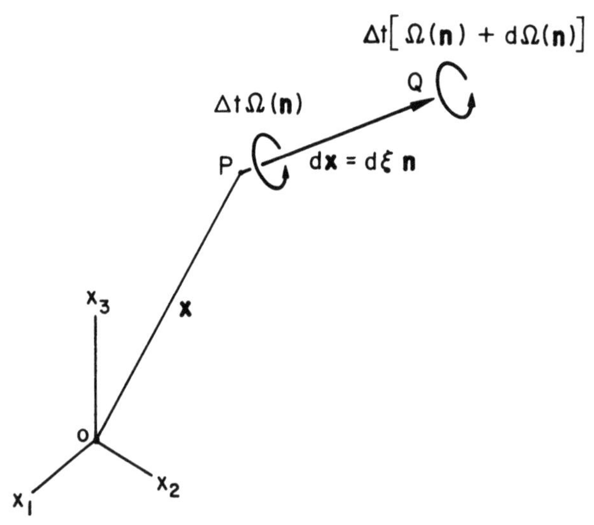

Fig. 1.6.1 Twisting of a line element.

terial line element PQ, $d\mathbf{x} = d\xi\mathbf{n}$, at the point $\mathbf{x}$ in the direction of the unit vector $\mathbf{n}$. In time Δt, the rotation of the element at P about $\mathbf{n}$ will be $\Delta t\ \Omega(\mathbf{n})$. However, this is a rotation only in the average sense described in Section (1.5). In the same interval, the rotation at Q will be $\Delta t[\Omega(\mathbf{n}) + d\Omega(\mathbf{n})]$. Thus the twist of Q relative to P will be $\Delta t\ d\Omega(\mathbf{n})$. The twist induced in the element PQ, per unit time, is therefore $d\Omega(\mathbf{n})$, so that the twist rate per unit length is given by $k(\mathbf{x},t;\mathbf{n}) = d\Omega(\mathbf{n})/d\xi$. Now $d\Omega(\mathbf{n})$ is the change in Ω over the distance $d\mathbf{x}$, so that $d\Omega = \Omega_{,r}dx_r$ and $d\Omega/d\xi = \Omega_{,r}n_r$, since $dx_r/d\xi = n_r$. Finally, a use of $\Omega(\mathbf{n}) = \omega_s n_s$, which implies that $\Omega_{,r} = \omega_{s,r}n_r$, gives

$$k(\mathbf{x},t;\mathbf{n}) = k_{rs}n_r n_s = \mathbf{n}^T\mathbf{K}\mathbf{n} \tag{1.6.1}$$

The second order pseudotensor $k_{rs} = \omega_{s,r}$, which is the vorticity gradient tensor, is called the curvature-twist rate tensor.

Next consider the curvature that is induced in line elements. Let $\mathbf{n}_1 = (l_i)$, $\mathbf{n}_2 = (m_i)$ and $\mathbf{n}_3 = (n_i)$ be a set of right-handed orthonormal vectors. Section (1.5) showed that the average rotation rate of any two line elements in the $\mathbf{n}_1 - \mathbf{n}_2$ plane about the $\mathbf{n}_3$ axis is given by $\Omega(\mathbf{n}_3) = \mathbf{n}_3^T\boldsymbol{\omega}$. A rotation of a line element in the $\mathbf{n}_1 - \mathbf{n}_2$ plane will mean this average rotation. With reference to Fig. 1.6.2, let $\overrightarrow{PQ} = d\mathbf{x} = d\xi\mathbf{n}_1$ be a material line

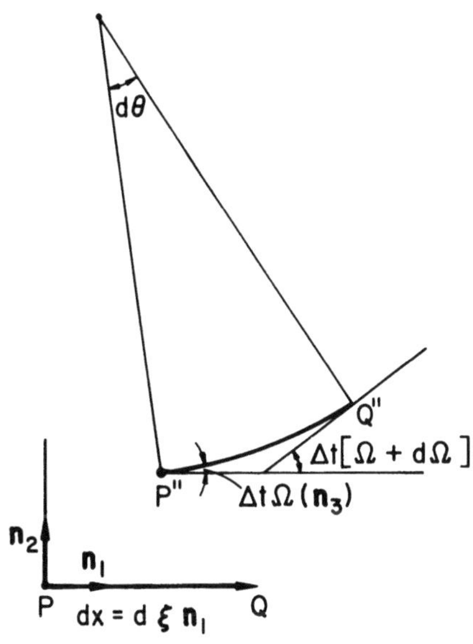

Fig. 1.6.2 Curvature induced in a line element.

element of length $d\xi$. After time Δt, P and Q will move to P' and Q', respectively. Let the projection of $P'Q'$ onto the $\mathbf{n}_1 - \mathbf{n}_2$ plane be $P''Q''$, as shown in Fig. 1.6.2. Then the rotation of the element will be such that the rotation at P'' and Q'', about $\mathbf{n}_3$, will be $\Delta t\, \Omega(\mathbf{n}_3)$ at P'' and $\Delta t[\Omega + d\Omega]$ at Q''. From the geometry shown in the figure, $d\theta = \Delta t\, d\Omega$. The curvature induced by $d\mathbf{x}$, in the $\mathbf{n}_1 - \mathbf{n}_2$ plane, is then given by $d\theta/d\xi = \Delta t\, d\Omega/d\xi$. Therefore, the rate at which curvature is induced per unit time is $d\Omega/d\xi$ and this is denoted by $k(\mathbf{x},t;\mathbf{n}_1,\mathbf{n}_3)$. But $d\Omega = \Omega_{,r}dx_r$, so that $d\Omega/d\xi = \Omega_{,r}dx_r/d\xi = \Omega_{,r}l_r$, and, since $\Omega = \omega_s n_s$,

$$k(\mathbf{x},t;\mathbf{n}_1,\mathbf{n}_3) = k_{rs}l_r n_s = \mathbf{n}_1^T\mathbf{K}\mathbf{n}_3 \tag{1.6.2}$$

Thus $k(\mathbf{x},t;\mathbf{n}_1,\mathbf{n}_3)$ is the rate at which curvature is induced, in the $\mathbf{n}_1 - \mathbf{n}_2$ plane, in a line element originally along $\mathbf{n}_1$, due to a variation of $\Omega(\mathbf{n}_3)$ along $\mathbf{n}_1$. For convenience, this will be abbreviated as $k(\mathbf{n}_1,\mathbf{n}_3)$. Thus both twist rates and the rates at which curvature is induced, are measured by $\mathbf{K}$. For this reason it is called the curvature-twist rate tensor.

The diagonal elements of $\mathbf{K}$ give the rates of twisting along the coordinate axes. For example, $k_{11} = \partial\omega_1/\partial x_1$ is the rate of twist of a line element along the x_1 axis. The off-diagonal elements measure the rate at which curvature is induced in line elements along the coordinate axes. For example, $k_{12} = \partial\omega_2/\partial x_1$ is the rate at which curvature is induced in the $x_1 - x_3$ plane, in a line element along the x_1 axis, due to a variation of ω_2 along x_1.

By definition $k_{ij} = \omega_{j,i} = \frac{1}{2}e_{jrs}v_{s,ri}$, so that $k_{ii} = \frac{1}{2}e_{irs}v_{s,ri} = 0$. Thus

$$k_{rr} = 0 \quad or \quad \mathrm{tr}\mathbf{K} = 0 \tag{1.6.3}$$

so that $\mathbf{K}$ has only eight independent components. The equation $\mathrm{tr}\mathbf{K} = 0$ may be interpreted as stating that the sum of the twist rates about any three mutually orthogonal directions is zero.

The curvature twist tensor was first discussed by Mindlin and Tiersten [Ref. 1], and Mindlin [Ref. 2], in connection with a theory of elasticity with couple stresses. Later, Stokes [Ref. 3] introduced the curvature-twist rate tensor for a theory of fluids with couple stresses.

1.7 Objective Tensors

A Cartesian tensor is defined as an object $a_{i_1 i_2 \cdots i_n}$, having 3^n components in a rectangular Cartesian frame Ox_i, with origin O, which when viewed from another Cartesian from $\overline{Ox}_i$, with origin $\overline{O}$, related to Ox_i by $\bar{x}_i = Q_{ir}x_r + c_i$, where $Q_{ir}Q_{jr} = \delta_{ij}$, is seen to have the components

$$\bar{a}_{i_1 i_2 \cdots i_n} = Q_{i_1 j_1}Q_{i_2 j_2} \cdots Q_{i_n j_n}a_{j_1 j_2 \cdots j_n} \tag{1.7.1}$$

In the coordinate transformation $\bar{x}_i = Q_{ir}x_r + c_i$, Q_{ir} and c_i are assumed to be independent of x_i and t. The transformation law given in Eq. (1.7.1) accounts for the fact that the tensor is being observed from two frames that are rotated relative to each other. Notice that the shift of the origin, represented by c_i, does not affect the transformation law.

Now consider the transformation of coordinates, given by $\bar{x}_i = Q_{ir}(t)x_r + c_i(t)$, where $Q_{ir}Q_{jr} = \delta_{ij}$, that relates two Cartesian frames that are moving relative to each other. Notice that Q_{ir} and c_i are now functions of time, but not of the coordinates. Because of this time dependence, many vectors and tensors will not follow the tensor-like transformation law. For example, a vector $\mathbf{v}$ may not transform as $\bar{v}_i = Q_{ir}v_r$ under this time-dependent coordinate transformation.

Since $\mathbf{x} = (x_i)$ and $\bar{\mathbf{x}} = (\bar{x}_i)$ are the position vectors of the same material point in two different Cartesian frames, the velocities, as observed by observers fixed to these frames, are given by $\mathbf{v} = d\mathbf{x}/dt$ and $\bar{\mathbf{v}} = d\bar{\mathbf{x}}/dt$. The relation between $\mathbf{v}$ and $\mathbf{v}$ may be obtained by differentiating the coordinate transformation $\bar{x}_i = Q_{ir}x_r + c_i$, which results in

$$\bar{v}_i = Q_{ir}v_r + \dot{Q}_{ir}x_r + \dot{c}_i \tag{1.7.2}$$

or $\bar{\mathbf{v}} = \mathbf{Q}\mathbf{v} + \dot{\mathbf{Q}}\mathbf{x} + \dot{\mathbf{c}}$, where a superposed dot indicates differentiation with respect to time. In addition to the term $\mathbf{Qv}$, which accounts for a rotation of the two frames relative to each other, there are terms $\dot{\mathbf{Q}}\mathbf{x} + \dot{\mathbf{c}}$. Thus the velocity does not transform as a vector if the two coordinate frames are moving relative to each other in time.

Next consider the behavior of the velocity gradient $g_{ij} = v_{j,i}$ under such a time-dependent transformation. From Eq. (1.7.2), since Q_{ij} and c_i are only functions of time,

$$\begin{aligned}\bar{g}_{ij} &= \frac{\partial \bar{v}_j}{\partial \bar{x}_i} = Q_{js}\frac{\partial v_s}{\partial \bar{x}_i} + \dot{Q}_{js}\frac{\partial x_s}{\partial \bar{x}_i} \\ &= Q_{js}v_{s,r}\frac{\partial x_r}{\partial \bar{x}_i} + \dot{Q}_{jr}\frac{\partial x_r}{\partial \bar{x}_i}\end{aligned}$$

or, since $\partial x_r/\partial \bar{x}_i = Q_{ir}$,

$$\bar{g}_{ij} = Q_{ir}Q_{js}g_{rs} + Q_{ir}\dot{Q}_{jr} \tag{1.7.3}$$

Thus the velocity gradient tensor also does not transform like a tensor under a time-dependent transformation. The term $Q_{ir}\dot{Q}_{jr}$ is skew-symmetric, for, a time differentiation of the orthogonality condition $Q_{ir}Q_{jr} = \delta_{ij}$ gives $Q_{ir}\dot{Q}_{jr} + \dot{Q}_{ir}Q_{jr} = 0$, so that

$$Q_{ir}\dot{Q}_{jr} = -Q_{jr}\dot{Q}_{ir} \tag{1.7.4}$$

Then, using Eqs. (1.7.3) and (1.7.4),

$$\bar{d}_{ij} = Q_{ir} Q_{js} d_{rs} \tag{1.7.5}$$

$$\bar{w}_{ij} = Q_{ir} Q_{js} w_{rs} + Q_{ir} \dot{Q}_{jr} \tag{1.7.6}$$

Thus, even though the velocity gradient tensor does not transform like a tensor under a time-dependent transformation, its symmetric part **D** does. However, the skew-symmetric part **W** also does not transform like a tensor.

Those tensors that preserve their transformation law even when the Cartesian frames are moving relative to each other, are called objective tensors. Thus the rate of deformation tensor **D** is objective, while the spin tensor **W** is not.

Using Eq. (1.7.3), the transformation law for the vorticity may be obtained as follows:

$$\begin{aligned}
2\bar{\omega}_j &= \bar{e}_{jkl} \bar{w}_{kl} = \bar{e}_{jkl}(Q_{kr} Q_{ls} w_{rs} + Q_{kr} \dot{Q}_{lr}) \\
&= (\det\mathbf{Q})\, Q_{jm} Q_{kn} Q_{lp} e_{mnp}(Q_{kr} Q_{ls} w_{rs} + Q_{kr} \dot{Q}_{lr}) \\
&= (\det\mathbf{Q})\,[Q_{jm} e_{mnp} w_{np} + Q_{jm} Q_{lp} e_{mnp} \dot{Q}_{ln}]
\end{aligned}$$

or

$$\bar{\omega}_j = (\det\mathbf{Q})\, Q_{jm} \omega_m + (\det \mathbf{Q})\, e_{mnp} Q_{jm} Q_{lp} \dot{Q}_{ln} \tag{1.7.7}$$

which shows that $\boldsymbol{\omega}$ is not objective. However,

$$\bar{\omega}_{j,i} = (\det\mathbf{Q})\, Q_{jn} Q_{im} \omega_{n,m}$$

or

$$\bar{k}_{ij} = (\det\mathbf{Q})\, Q_{ir} Q_{jr} k_{rs} \tag{1.7.8}$$

thereby showing that the curvature-twist rate tensor is an objective pseudotensor.

Only objective quantities can be proper measures of deformation rates, as the effects of nonobjective terms can be transformed away by an appropriate choice of the coordinate frame.

1.8 Balance of Mass

Consider the flow of a continuous medium, such as a fluid, through a fixed region of space V, bounded by the surface ∂V. If ρ is the mass density of the fluid, then the total mass of fluid in the region V, at time t, will be $\int_V \rho \, dV$, so that the rate of increase of mass will be

$\frac{\partial}{\partial t}\int_V \rho\, dV = \int_V \frac{\partial \rho}{\partial t}\, dV$. If $\mathbf{v}(\mathbf{x},t)$ is the velocity at the point $\mathbf{x}$, at time t, then the rate at which fluid is coming into the region V, through an element of area dA of ∂V having normal $\mathbf{n}$, must be $-\rho v_r n_r\, dA$, so that the total influx rate of fluid into V is $-\int_{\partial V} \rho v_r n_r\, dA$, which upon using Gauss' theorem, becomes $-\int_{\partial V} (\rho v_r)_{,r}\, dV$. Now the balance principle for mass, also known as conservation of mass, states that the rate of increase of the mass inside a region of space V, must equal the rate at which mass is coming into V through the boundary ∂V, so that $\int_V (\partial \rho/\partial t)\, dV = -\int_V (\rho v_r)_{,r}\, dV$ or $\int_V [\partial \rho/\partial t + (\rho v_r)_{,r}]\, dV = 0$. Since this must hold for arbitrary volume V, the balance of mass is governed by the continuity equation

$$\frac{\partial \rho}{\partial t} + (\rho v_r)_{,r} = 0 \quad or \quad \dot{\rho} + \rho v_{r,r} = 0 \tag{1.8.1}$$

where a superposed dot indicates the material time derivative $D/Dt = \partial/\partial t + v_r \partial/\partial x_r$. This, of course, is the continuity equation. For an incompressible fluid $\dot{\rho} = D\rho/Dt = 0$, so that the continuity equation becomes $v_{r,r} = 0$, which is the same as $d_{rr} = 0$ or $\mathrm{tr}\mathbf{D} = 0$, or even $\mathrm{tr}\mathbf{G} = 0$.

1.9 Concluding Remarks

This chapter has shown that the rate of deformation tensor **D**, which is used in fluid mechanics, does not give a complete picture of deformation rates, but only gives a measure of the time rates of normal and shear strains. For example, it does not give any information about the rates of twisting or the rates at which curvature is induced in line elements.

In addition to the time rates of normal and shear strain, the time rates of twisting and induction of curvature might also be expected to affect the state of stress. However, in Newtonian fluids, of the type in which the constitutive equation has the form $\mathbf{T} = -p\mathbf{I} + \lambda(\mathrm{tr}\mathbf{D})\mathbf{I} + 2\mu\mathbf{D}$, the stress is assumed to be only affected by **D** and not by **K**. The entire structure of classical fluid mechanics of viscous flow is based on the assumption that deformation rates are completely measured by **D**.

Of course, the introduction of **K** does not mean that the objective tensors **D** and **K** completely determine all aspects of deformation rates. Note that whereas **D** is made up of velocity gradients, the pseudotensor **K** is made up of second order gradients of the velocity. In this sense **K** is a higher order measure for deformation rates.

1.10 References

1. Mindlin, R. D., and Tiersten, H. F. (1962). Effects of Couple Stresses in Linear Elasticity, *Arch. Ration. Mech. Anal.* **11**, 415-448.
2. Mindlin, R. D. (1963). Influence of Couple Stresses in Linear Elasticity, *Exp. Mech.* **20**, 1-7.
3. Stokes, V. K. (1966). Couple Stresses in Fluids, *Phys. Fluids* **9**, 1709-1715.

Chapter 2

Field Equations

2.1 Introduction

This chapter reviews the field equations of continuum mechanics that are relevant to theories of fluids with microstructure. The emphasis is on explaining the significance of couple stresses and their effects on the field equations. The discussion is confined to a study of continuous media.

2.2 Measures for Mechanical Interactions

Mechanical interactions are those that are measured by forces and moments. The general problem of continuum mechanics is to determine the effects of forces and moments on the motion of a continuous body, and it requires the establishment of measures for the mechanical interactions that occur inside a deforming continuum. Consider a continuous body B_o, bounded by the surface ∂B_o, which is in motion under the action of forces and moments, as shown in Fig. 2.2.1. In order to define the mechanical in-

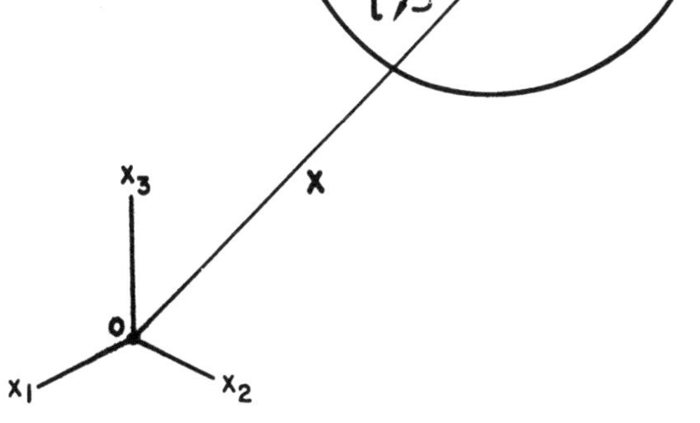

Fig. 2.2.1 A body B_o bounded by the surface ∂B_o.

teractions within such a body, consider an arbitrary portion B of B_o, bounded by the surface ∂B. The mechanical effect of $(B_o - B)$ on B will be through forces and moments distributed over the surface ∂B, and since they vary over the surface, area densities of these forces and moments are more appropriate measures. $\mathbf{t}(\mathbf{x},t;\mathbf{n})$ is then defined to be the force per unit area acting at the point $\mathbf{x}$, at time t, on a unit area of orientation $\mathbf{n}$, so that the force on an infinitesimal area dA is $\mathbf{t}dA$. Similarly $\mathbf{m}(\mathbf{x},t;\mathbf{n})$ is defined as the couple per unit area at $(\mathbf{x},t)$ on an area with orientation $\mathbf{n}$, such that $\mathbf{m}dA$ gives the moment acting on $\mathbf{n}dA$. $\mathbf{t}(\mathbf{x},t;\mathbf{n})$ and $\mathbf{m}(\mathbf{x},t;\mathbf{n})$ are called the stress vector and the couple stress vector respectively. Their vectorial nature follows from the known behavior of forces and moments. Since forces are vectors, so also will be the stress vector. However, since moments are pseudovectors, the couple stress vector is really a pseudovector.

Thus the fields $\mathbf{t}(\mathbf{x},t;\mathbf{n})$ and $\mathbf{m}(\mathbf{x},t;\mathbf{n})$ distributed over ∂B describe, or represent, the entire mechanical effect of $(B_o - B)$ on B. In effect, the body $\mathbf{B}$ will not sense the removal of $(B_o - B)$ if ∂B is acted upon by the fields $\mathbf{t}(\mathbf{x},t;\mathbf{n})$ and $\mathbf{m}(\mathbf{x},t;\mathbf{n})$, whose action is therefore equivalent to that of $(B_o - B)$.

The sum of all the forces acting on the surface ∂B is called the contact force $\mathbf{F}_c$, and is given by

$$\mathbf{F}_c = \int_{\partial B} \mathbf{t}\, dA \tag{2.2.1}$$

Similarly, the sum of all the torques acting on ∂B, about the origin, is called the contact moment, and it is given by

$$\mathbf{L}_c = \int_{\partial B} \mathbf{m}\, dA + \int_{\partial B} \mathbf{x}\times\mathbf{t}\, dA \tag{2.2.2}$$

Notice that the stress vector contributes to the contact torque.

In addition to contact forces and moments that represent the action of $(B_o - B)$ on B, the body B may also be subjected to body forces and body moments, which result from fields that operate directly on B. Let $\mathbf{f}(\mathbf{x},t)$ and $\mathbf{l}(\mathbf{x},t)$ be, respectively, the body force per unit mass and the body moment per unit mass acting on B. Then the total body force $\mathbf{F}_b$ and the total body moment $\mathbf{L}_b$, acting on B, are given by

$$\mathbf{F}_b = \int_B \rho \mathbf{f}\, dV \tag{2.2.3}$$

$$\mathbf{L}_b = \int_B \rho \mathbf{l}\, dV + \int_B \rho \mathbf{x}\times\mathbf{f}\, dV \tag{2.2.4}$$

For example, a body placed in a gravitational field would be subjected to a body force. Similarly a magnetic body in a magnetic field would be subjected to a body moment.

The sums of all the forces, $\mathbf{F}$, and all the moments about the origin, $\mathbf{L}$, acting on B, are then given by

$$\mathbf{F} = \mathbf{F}_c + \mathbf{F}_b = \int_{\partial B} \mathbf{t} \, dA + \int_B \rho \mathbf{f} \, dV \tag{2.2.5}$$

$$\mathbf{L} = \mathbf{L}_c + \mathbf{L}_b = \int_{\partial B} \mathbf{m} \, dA + \int_{\partial B} \mathbf{x} \times \mathbf{t} \, dA + \int_B \rho \mathbf{l} \, dV + \int_B \rho \mathbf{x} \times \mathbf{f} \, dV \tag{2.2.6}$$

Once the total force and the total moment acting on B are known, they may be used for obtaining the equations of motion by using Euler's two laws of motion.

Unfortunately, the area densities of the surface forces and moments, $\mathbf{t}(\mathbf{x},t;\mathbf{n})$ and $\mathbf{m}(\mathbf{x},t;\mathbf{n})$, respectively, are not point functions because they depend on the orientation vector $\mathbf{n}$. The stress and couple stress tensors are introduced in order to obtain measures that are point functions.

2.3 Euler's Laws of Motion

The laws of motion used are those due to Euler. Euler's first law of motion states that the time rate of change of the total linear momentum, $\mathbf{P}$, of a body is equal to the total force $\mathbf{F}$ acting on the body. Euler's second law states that the time rate of change of the total angular momentum, $\mathbf{H}$, of the body about any point, equals the total torque, $\mathbf{L}$, acting on the body about that point. That is

$$\dot{\mathbf{P}} = \mathbf{F} \tag{2.3.1}$$

$$\dot{\mathbf{H}} = \mathbf{L} \tag{2.3.2}$$

where the superposed dot indicates the material time derivative D/Dt.

The total linear momentum $\mathbf{P}$ and the angular momentum $\mathbf{H}$ may be expressed in terms of their densities per unit mass, $\mathbf{p}$ and $\mathbf{h}$, respectively, by

$$\mathbf{P} = \int_B \rho \mathbf{p} \, dV \tag{2.3.3}$$

$$\mathbf{H} = \int_B \rho \mathbf{h} \, dV \tag{2.3.4}$$

In classical continuum mechanics small volume elements are assumed to have no microstructure, in the sense that the kinematics of motion is completely specified by the velocity of each point, and the linear and angular momenta are specified by the densities

$$\mathbf{p} = \mathbf{v} \quad , \quad \mathbf{h} = \mathbf{x} \times \mathbf{v} = \mathbf{x} \times \mathbf{p} \tag{2.3.5}$$

where, for convenience, $\mathbf{H}$ and $\mathbf{L}$ are measured about the origin. Then, in the absence of the velocity field, both $\mathbf{p}$ and $\mathbf{h}$, and hence $\mathbf{P}$ and $\mathbf{H}$, are zero.

If a continuum is considered to consist of a large assemblage of small individual particles, then there is the possibility of an angular momentum even in the absence of a macroscopic field. For, in such a model, even though particles may not have a macroscopic velocity, they may be spinning individually, each particle thereby having an intrinsic angular momentum. This effect can be accounted for by introducing an intrinsic angular momentum density per unit mass, $\boldsymbol{\sigma}$, such that the angular momentum is specified by $\mathbf{h} = \mathbf{x}\times\mathbf{v} + \boldsymbol{\sigma}$. Of course, this is only possible if the fluid is assumed to have a microstructure, so that the macroscopic field does not specify the motion completely. If such a spin momentum is taken into account, then the linear and angular momenta are specified by

$$\mathbf{p} = \mathbf{v} \quad , \quad \mathbf{h} = \mathbf{x}\times\mathbf{v} + \boldsymbol{\sigma} = \mathbf{x}\times\mathbf{p} + \boldsymbol{\sigma} \tag{2.3.6}$$

Thus, the linear momentum will always be specified by $\mathbf{p} = \mathbf{v}$. However, for the angular momentum, the term $\boldsymbol{\sigma}$ in $\mathbf{h} = \mathbf{x}\times\mathbf{v} + \boldsymbol{\sigma}$ may or may not be zero, depending on whether microstructure is not, or is, taken into account.

Over the body B, the density ρ may change with time but, because of conservation of mass $\rho\, dV$, the mass of each element, remains a constant, so that if $\psi(\mathbf{x},t)$ is any property associated with the fluid, then

$$\frac{D}{Dt}\int_B \rho\psi(\mathbf{x},t)\, dV = \int_B \rho\dot{\psi}\, dV \tag{2.3.7}$$

The rate of change of linear and angular momenta of B are then given by

$$\dot{\mathbf{P}} = \int_B \rho\dot{\mathbf{p}}\, dV \tag{2.3.8}$$

$$\dot{\mathbf{H}} = \int_B \rho\dot{\mathbf{h}}\, dV \tag{2.3.9}$$

where $\dot{\mathbf{p}} = \mathbf{a}$, $\dot{\mathbf{h}} = \mathbf{x}\times\dot{\mathbf{p}} + \dot{\boldsymbol{\sigma}}$, and $\mathbf{a} = \dot{\mathbf{v}} = \partial\mathbf{v}/\partial t + \mathbf{v}^T\nabla\mathbf{v}$ is the acceleration at $\mathbf{x}$ at time t.

2.4 *Stress and Couple Stress Vectors*

The stress and couple stress vectors satisfy

$$\mathbf{t}(\mathbf{x},t;-\mathbf{n}) = -\mathbf{t}(\mathbf{x},t;\mathbf{n}) \tag{2.4.1}$$

$$\mathbf{m}(\mathbf{x},t;-\mathbf{n}) = -\mathbf{m}(\mathbf{x},t;\mathbf{n}) \tag{2.4.2}$$

For, at $(\mathbf{x},t)$, consider the motion of a thin material disk of area ΔA and uniform thickness dn, which has the unit normal $\mathbf{n}$, as shown in Fig. 2.4.1.

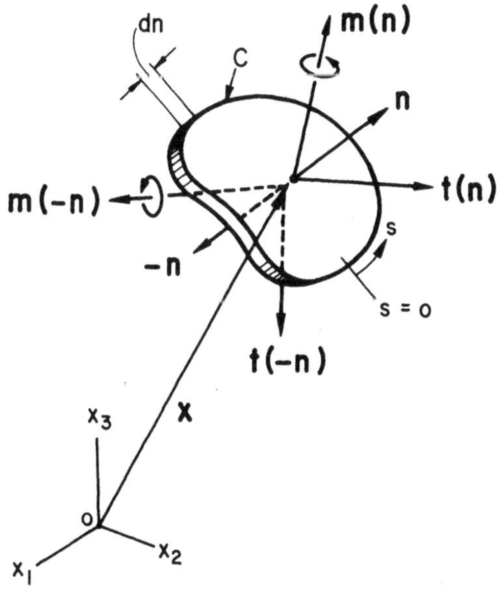

Fig. 2.4.1 Stresses acting on a thin disk.

Then, by definition, the stress and couple stress vectors will be $\mathbf{t}(\mathbf{x},\mathbf{n})$, $\mathbf{m}(\mathbf{x},\mathbf{n})$ and $\mathbf{t}(\mathbf{x},-\mathbf{n})$, $\mathbf{m}(\mathbf{x},-\mathbf{n})$ on the sides with normals $\mathbf{n}$ and $-\mathbf{n}$, respectively. The total force and moment on the disk are given by

$$\mathbf{F} = \Delta A\,[\mathbf{t}(\mathbf{n}) + \mathbf{t}(-\mathbf{n})] + dn\int_C \mathbf{t}(s)\;ds + dn\;\Delta A\;\rho\mathbf{f} \tag{2.4.3}$$

$$\mathbf{H} = \Delta A\,[\mathbf{m}(\mathbf{n}) + \mathbf{m}(-\mathbf{n})] + \Delta A\;\mathbf{x}\times[\mathbf{t}(\mathbf{n}) + \mathbf{t}(-\mathbf{n})] + dn\int_C \mathbf{m}(s)\;ds$$
$$+\; dn\int_C \mathbf{x}\times\mathbf{t}(s)\;ds + dn\;\Delta A\,[\rho\mathbf{l} + \rho\mathbf{x}\times\mathbf{f}] \tag{2.4.4}$$

where the term $dn\int_C \mathbf{t}(s)\;ds$ in Eq. (2.4.3) is the contribution to the contact force due to the stress vector $\mathbf{t}(s)$ acting at a distance s on the edge C of the disk, and so on. The linear and angular momenta of the disk are $\mathbf{P} = dn\;\Delta A\;\rho\mathbf{p}$ and $\mathbf{H} = dn\;\Delta A\,[\rho\mathbf{x}\times\mathbf{p} + \rho\boldsymbol{\sigma}]$, respectively, so that $\dot{\mathbf{P}} = dn\;\Delta A\;\dot{\mathbf{p}}$ and $\dot{\mathbf{H}} = dn\;\Delta A\,[\rho\mathbf{x}\times\dot{\mathbf{p}} + \rho\dot{\boldsymbol{\sigma}}]$, as $\dot{\mathbf{x}}\times\mathbf{p} = \mathbf{p}\times\mathbf{p} = 0$. Use of Euler's laws of motion, as given by Eqs. (2.3.1) and (2.3.2), then results in

$$dn\,\Delta A\,\rho\dot{\mathbf{p}} = \Delta A\,[\mathbf{t}(\mathbf{n}) + \mathbf{t}(-\mathbf{n})] + dn\int_C \mathbf{t}(s)\,ds + dn\,\Delta A\,\rho\mathbf{f}$$

$$dn\,\Delta A\,[\rho\dot{\boldsymbol{\sigma}} + \rho\mathbf{x}\times\dot{\mathbf{p}}] = \Delta A\,[\mathbf{m}(\mathbf{n}) + \mathbf{m}(-\mathbf{n})] + dn\int_C \mathbf{m}(s)\,ds$$

$$+dn\int_C \mathbf{x}\times\mathbf{t}(s)\,ds + \Delta A\;\mathbf{x}\times[\mathbf{t}(\mathbf{n}) + \mathbf{t}(-\mathbf{n})]$$

$$+\,dn\,\Delta A\,[\rho\mathbf{l} + \rho\mathbf{x}\times\mathbf{f}]$$

Taking the limit $dn \rightarrow 0$, these two equations reduce to $\Delta A\,[\mathbf{t}(\mathbf{n}) + \mathbf{t}(-\mathbf{n})] = 0$ and $\Delta A\,[\mathbf{m}(\mathbf{n}) + \mathbf{m}(-\mathbf{n})] + \Delta A\;\mathbf{x}\times[\mathbf{t}(\mathbf{n}) + \mathbf{t}(-\mathbf{n})] = 0$, so that $\mathbf{t}(\mathbf{n}) + \mathbf{t}(-\mathbf{n}) = 0$, and $\mathbf{m}(\mathbf{n}) + \mathbf{m}(-\mathbf{n}) + \mathbf{x}\times[\mathbf{t}(\mathbf{n}) + \mathbf{t}(-\mathbf{n})] = 0$, which, together, give the desired results. Notice that $\mathbf{t}(-\mathbf{n}) = -\mathbf{t}(\mathbf{n})$ is a direct consequence of Euler's first law of motion. On the other hand, an application of the second law gives $\mathbf{m}(\mathbf{n}) + \mathbf{m}(-\mathbf{n}) + \mathbf{x}\times[\mathbf{t}(\mathbf{n}) + \mathbf{t}(-\mathbf{n})] = 0$, so that both the first and second laws are required for establishing $\mathbf{m}(\mathbf{n}) + \mathbf{m}(-\mathbf{n}) = 0$.

The results in Eqs. (2.4.1) and (2.4.2) may be referred to as Cauchy's lemmas, and will be used in establishing the stress and couple stress tensors.

2.5 *Stress and Couple Stress Tensors*

The next step is to establish measures for mechanical interactions, so far described by the orientation dependent vectors $\mathbf{t}$ and $\mathbf{m}$, which are point functions. The main results are Cauchy's fundamental theorem and its generalization, which establish the existence of second order tensors $\mathbf{T} = (t_{ij})$ and $\mathbf{M} = (m_{ij})$, called the stress and couple stress tensors, respectively, through

$$t_i = t_{ri}\,n_r \quad or \quad \mathbf{t}(\mathbf{x},t;\mathbf{n}) = \mathbf{T}^T(\mathbf{x},t)\,\mathbf{n} \tag{2.5.1}$$

$$m_i = m_{ri}\,n_r \quad or \quad \mathbf{m}(\mathbf{x},t;\mathbf{n}) = \mathbf{M}^T(\mathbf{x},t)\,\mathbf{n} \tag{2.5.2}$$

The tensors $\mathbf{T}$ and $\mathbf{M}$, each having nine components, then describe the mechanical interactions, or the state of stress, at a point. In effect they factor out the dependence of $\mathbf{t}$ and $\mathbf{m}$ on $\mathbf{n}$.

By choosing $\mathbf{n} = \mathbf{i}_1 = (1\;0\;0)$, the components of $\mathbf{t}(\mathbf{i}_1)$ are given by $\mathbf{t}(\mathbf{i}_1) = (t_{11}\;t_{12}\;t_{13})$. Thus, if $\mathbf{T}$ and $\mathbf{M}$ exist, then t_{ij} gives the *jth* component of the stress vector on the coordinate plane defined by x_i = constant. t_{1r}, $r = 1, 2, 3$, gives the three components of the stress vector on the plane x_1 = constant, and so on. Similarly m_{1r}, $r = 1, 2, 3$, are the three components of the couple stress vector on the plane x_1 = constant. $\mathbf{T}$ would therefore seem to be determined once $\mathbf{t}$ is known on three orthogonal planes. The formal results will now be enunciated and proved.

Cauchy's Fundamental Theorem states that if the stress vector is known on three mutually orthogonal planes through a point, then it can be determined on any plane passing through that point. Furthermore, $\mathbf{t}(\mathbf{x},\mathbf{n}) = \mathbf{T}^T(\mathbf{x})\mathbf{n}$, where $\mathbf{T}$ is the matrix whose first, second, and third rows are the components of the stress vectors on the three given planes. In terms of Cartesian tensors this result is written as $t_i = t_{ji} n_j$.

In order to prove this result, choose a Cartesian coordinate system in which the coordinate planes are parallel to the three orthogonal planes on which the stress vector is known. At the point P, $(\mathbf{x},t)$, choose an infinitesimal tetrahedron $PABC$, in which the plane ABC, with normal $\mathbf{n}$, is at a distance Δs from PE, as shown in Fig. 2.5.1. Let PA, PB and PC be

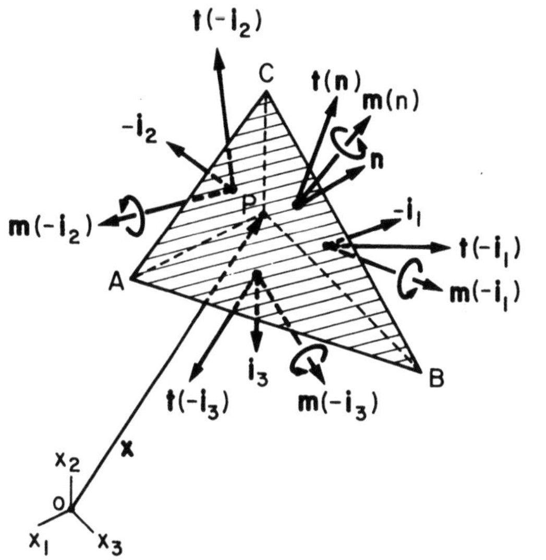

Fig. 2.5.1 Stresses acting on an infinitesimal tetrahedron.

Δx_1, Δx_2 and Δx_3, respectively. If the area of ABC is ΔA, then the areas BPC, CPA and APB, being the projections of ΔA on the coordinate planes, are given by $\Delta A_{x_1} = n_1 \Delta A$, $\Delta A_{x_2} = n_2 \Delta A$ and $\Delta A_{x_3} = n_3 \Delta A$, respectively. By definition, the contact force on the tetrahedron is then given by

$$\mathbf{F}_c = \Delta A \ \mathbf{t}(\mathbf{n}) + \Delta A_{x_1}\mathbf{t}(-\mathbf{i}_1) + \Delta A_{x_2}\mathbf{t}(-\mathbf{i}_2) + \Delta A_{x_3}\mathbf{t}(-\mathbf{i}_3)$$

which, upon using Cauchy's lemma, and $\Delta A_{x_i} = n_i \, \Delta A$, reduces to

$$\mathbf{F}_c = \Delta A \left[\mathbf{t}(\mathbf{n}) - n_1\mathbf{t}(\mathbf{i}_1) - n_2\mathbf{t}(\mathbf{i}_2) - n_3\mathbf{t}(\mathbf{i}_3)\right] \tag{2.5.3}$$

Similarly, the contact moment is given by

$$\mathbf{L}_c = \Delta A\,[\mathbf{m}(\mathbf{n}) - n_1\mathbf{m}(\mathbf{i}_1) - n_2\mathbf{m}(\mathbf{i}_2) - n_3\mathbf{m}(\mathbf{i}_3)] + \Delta A\ \mathbf{x}\times[\mathbf{t}(\mathbf{n}) - n_1\mathbf{t}(\mathbf{i}_1) - n_2\mathbf{t}(\mathbf{i}_2) - n_3\mathbf{t}(\mathbf{i}_3)] \tag{2.5.4}$$

The body force and body moment acting on the tetrahedron will be

$$\mathbf{F}_b = \Delta V\,\rho\mathbf{f}\quad,\quad \mathbf{L}_b = \Delta V\,(\rho\mathbf{l} + \rho\mathbf{x}\times\mathbf{f}) \tag{2.5.5}$$

where $\Delta V = \frac{1}{3}\,\Delta s\,\Delta A$ is the volume of the tetrahedron. Since the time rates of change of linear and angular momenta of the tetrahedron are $\rho\,\Delta V\,\mathbf{a}$ and $\rho\,\Delta V(\dot{\boldsymbol{\sigma}} + \mathbf{x}\times\mathbf{a})$, respectively, Euler's two laws of motion, when applied to the tetrahedron, will give

$$\frac{1}{3}\,\Delta s\,\Delta A\,\rho\mathbf{a} = \Delta A\left[\mathbf{t}(\mathbf{n}) - \sum_{r=1}^{3} n_r\mathbf{t}(\mathbf{i}_r)\right] + \frac{1}{3}\,\Delta s\,\Delta A\,\rho\mathbf{f}$$

$$\frac{1}{3}\,\Delta s\,\Delta A\,\rho(\dot{\boldsymbol{\sigma}} + \mathbf{x}\times\mathbf{a}) = \Delta A\left[\mathbf{m}(\mathbf{n}) - \sum_{r=1}^{3} n_r\mathbf{m}(\mathbf{i}_r)\right] + \Delta A\ \mathbf{x}\times\left[\mathbf{t}(\mathbf{n}) - \sum_{r=1}^{3} n_r\mathbf{t}(\mathbf{i}_r)\right] + \frac{1}{3}\,\Delta s\,\Delta A\,\rho(\mathbf{l} + \mathbf{x}\times\mathbf{f})$$

A division of these two equations by ΔA, followed by taking the limit of $\Delta s \to 0$, then results in

$$\mathbf{t}(\mathbf{n}) = n_i\mathbf{t}(\mathbf{i}_1) + n_2\mathbf{t}(\mathbf{i}_2) + n_3\mathbf{t}(\mathbf{i}_3) \tag{2.5.6}$$

$$\mathbf{m}(\mathbf{n}) = n_1\mathbf{m}(\mathbf{i}_1) + n_2\mathbf{m}(\mathbf{i}_2) + n_3\mathbf{m}(\mathbf{i}_3) - \mathbf{x}\times\left[\mathbf{t}(\mathbf{n}) - \sum_{r=1}^{3} n_r\mathbf{t}(\mathbf{i}_r)\right] \tag{2.5.7}$$

Finally, use of Eq. (2.5.6) reduces Eq. (2.5.7) to

$$\mathbf{m}(\mathbf{n}) = \sum_{r=1}^{3} n_r\mathbf{m}(\mathbf{i}_r) \tag{2.5.8}$$

Thus **t** has been shown to be determined on all planes at a point, once it is known on three mutually orthogonal planes at that point. The same is true of $\mathbf{m}$. Let t_{rs} be the component of $\mathbf{t}(\mathbf{i}_r)$ in the s direction, that is, let $\mathbf{t}(\mathbf{i}_r) = t_{rs}\mathbf{i}_s$. Then Eq. (2.5.6) can be written as $t_i = n_1 t_{1i} + n_2 t_{2i} + n_3 t_{3i} = n_r t_{ri}$, or $t_i = n_r t_{ri}$. Similarly, if m_{rs} are the components of $\mathbf{m}(\mathbf{i}_r)$ along the s direction, then $m_i = n_r m_{ri}$. Since $\mathbf{t}$ is known to be vector and since Eq. (2.5.1) holds for arbitrary $\mathbf{n}$, $\mathbf{T}$ is a second order tensor. Similarly, since $\mathbf{m}$ is a pseudovector, $\mathbf{M}$ is a second order pseudotensor. This completes the derivation of the results given in Eqs. (2.5.1) and (2.5.2).

Thus, Euler's laws of motion make it possible to define measures for mechanical interactions that are point functions. Note that whereas the relation $t_i = t_{ri} n_r$ is a consequence of Euler's first law, the relation $m_i = m_{ri} n_r$ can be obtained only by using both the first and second laws of Euler. The second law alone implies Eq. (2.5.7), that is, $m_i = m_{ri} n_r - e_{ijk} x_j [t_k - t_{rk} n_r]$.

2.6 Cauchy's Laws of Motion

Cauchy's laws of motion are obtained by applying Euler's laws of motion to a continuum. Consider then the motion of the body B_o, shown in Fig. 2.2.1. Using Eqs. (2.2.5) and (2.2.6), the total force F_i and the total moment L_i acting on an arbitrary portion B of B_o are then given by

$$F_i = \int_{\partial B} t_i \, dA + \int_B \rho f_i \, dV$$

$$L_i = \int_{\partial B} m_i \, dA + \int_{\partial B} e_{ijk} x_j t_k \, dA + \int_B \rho l_i \, dV + \int_B \rho \, e_{ijk} x_j f_k \, dV$$

By making use of Cauchy's fundamental theorem, $t_i = t_{ri} n_r$ and $m_i = m_{ri} n_r$, and applying the generalized Gauss' theorem in the form $\int_{\partial B} (\ ) n_r \, dA = \int_B (\ )_{,r} \, dV$, the surface integrals over ∂B may be converted into volume integrals over B as follows:

$$\int_{\partial B} t_i \, dA = \int_{\partial B} t_{ri} n_r \, dA = \int_B t_{ri,r} \, dV$$

$$\int_{\partial B} m_i \, dA = \int_{\partial B} m_{ri} n_r \, dA = \int_B m_{ri,r} \, dV$$

$$\int_{\partial B} e_{ijk} x_j t_k \, dA = \int_{\partial B} e_{ijk} x_j t_{rk} n_r \, dA = \int_B (e_{ijk} x_j t_{rk})_{,r} \, dV$$

$$= \int_B [e_{ijk} t_{jk} + e_{ijk} x_j t_{rk,r}] \, dV$$

since $x_{j,r} = \delta_{rj}$. The total force and moment acting on B are then

$$F_i = \int_B (t_{ri,r} + \rho f_i)\, dV \tag{2.6.1}$$

$$L_i = \int_B [m_{ri,r} + \rho l_i + e_{ijk} t_{jk} + e_{ijk} x_j (t_{rk,r} + \rho f_k)]\, dV \tag{2.6.2}$$

Since Eqs. (2.3.8) and (2.3.9) give the time rates of the linear and angular momenta of B as $\dot{P}_i = \int_B \rho a_i \, dV$ and $\dot{H}_i = \int_B (\rho \dot{\sigma}_i + \rho e_{ijk} x_j a_k)\, dV$, Euler's laws when applied to B give

$$\int_B \rho a_i \, dV = \int_B (t_{ri,r} + \rho f_i)\, dV$$

$$\int_B (\rho \dot{\sigma}_i + \rho e_{ijk} x_j a_k)\, dV = \int_B [m_{ri,r} + \rho l_i + e_{ijk} t_{jk} + e_{ijk} x_j (t_{rk,r} + \rho f_k)]\, dV$$

Furthermore, since these relations hold for all parts B of B_o, $\rho a_i = t_{ri,r} + \rho f_i$, and

$$\rho \dot{\sigma}_i + e_{ijk} x_j \left[\rho a_k - t_{rk,r} - \rho f_k\right] = m_{ri,r} + \rho l_i + e_{ijk} t_{jk} \tag{2.6.3}$$

Equation (2.6.3) may be simplified by eliminating the second term on the left-hand side, which must vanish by virtue of the first equation given above. Thus, the laws of motion have the form

$$\rho a_i = t_{ri,r} + \rho f_i \tag{2.6.4}$$

$$\rho \dot{\sigma}_i = m_{ri,r} + \rho l_i + e_{ijk} t_{jk} \tag{2.6.5}$$

These equations are called Cauchy's first and second laws of motion. Here again, while Eq. (2.6.4) follows directly from Euler's first law, Eq. (2.6.5) is a consequence of using both the first and second laws of Euler.

If internal spin, couple stresses, and body moments are zero, then Cauchy's second law implies that the axial vector $T_i^A = e_{ijk} t_{jk}$, associated with the stress tensor, vanishes, which is only possible if the skew-symmetric part of the stress tensor is zero. For example, $T_1^A = e_{1jk} = t_{23} - t_{32} = 0$ implies that $t_{23} = t_{32}$, and so on. The situation in which the internal spin, or the couple stresses, or the body moments are not zero, is referred to as the polar case. In the nonpolar case all these quantities are assumed to be zero, so that Cauchy's second law reduces to the statement that the stress tensor is symmetric.

Thus, in the nonpolar case, mechanical interactions are measured by the stress tensor alone which, because of symmetry, has only six independent components. Cauchy's first law of motion provides three equations in these six unknowns. On the other hand, in the polar case, mechanical interactions are measured by both **T** and **M**, neither of which is necessarily symmetric, so that the number of unknowns is eighteen. Cauchy's first and second laws provide only six equations in these unknowns. This discussion indicates the complexity that may be expected in a polar theory.

There is an important difference between stresses and couple stresses. For a given acceleration field and a body force distribution, Cauchy's law, given in Eq. (2.6.4), shows that the stress tensor must exist. That is, the equations of motion cannot be satisfied by assuming that the stresses are identically zero. On the other hand, the second law, represented by Eq. (2.6.5), can be satisfied by assuming that the couple stresses are zero.

In most analyses of mechanics, the mechanical interactions of one part of a body with its neighboring parts, through the common interface, are assumed to be equivalent to a force distribution alone. This assumption does not seem reasonable when even the laws of motion, in the form due to Euler, are expressed in terms of forces and moments.

2.7 Analysis of Stress

Analysis of stress is concerned with the following three questions:

(*a*) At any given point in a material, are there planes on which the shear stress is zero? Such planes, if they exist, are called the principal planes of stress and their normals the principal directions of stress. The (normal) stresses on these planes are called principal stresses.

(*b*) On which planes is the normal stress a maximum? And

(*c*) On which planes is the shear stress a maximum?

A. Principal Planes

On a plane n_i the stress vector is given by $t_i = t_{ri} n_r$, so that if n_i is a principal direction then t_i is parallel to n_i, that is $t_i = \sigma n_i$, say. The principal directions n_i are then the solutions of the equation $t_{ri} n_r = \sigma n_i$ or $\mathbf{T}^T \mathbf{n} = \sigma \mathbf{n}$. Thus the principal directions are the proper vectors of $\mathbf{T}^T$ and the principal stresses are the corresponding proper values.

In the nonpolar case, $\mathbf{T}$ is symmetric so that all the three proper values are real. A triple of orthogonal proper vectors, and hence principal directions, always exists. There is only one such triple if the three proper values are distinct. If two of the proper values are equal, then the direction corresponding to the distinct proper value and all directions normal to it will be principal directions. If all the proper values are equal, then every direction is a principal direction.

However, in the polar case the stress tensor is not symmetric, in general, so that all the proper values need not be real. Of course, one proper value is always real, so that at least one principal direction always exists. In fact for the polar case one, two, three, and an infinity of principal directions are possible. Even if three principal directions exist, they need not be orthogonal.

B. Maximum Normal Stress

In the nonpolar case the extreme values of the normal stress occur along the principal directions. That is, the principal stresses also give the extreme values of the normal stress. In the polar case, the extreme values of the normal stress occur along the proper vectors of the symmetric part $\mathbf{T}^S$ of $\mathbf{T}$, and the extreme values are given by the corresponding proper values of $\mathbf{T}^S$.

C. Maximum Shear Stress

For the nonpolar case, the extreme values of the shear stress occur on planes whose normals bisect the angles between the proper vectors of $\mathbf{T}$. However, no such simple characterization is available for the polar case.

Some aspects of the analysis of stress for the polar case are given in Refs. [1] and [2].

2.8 Energy Balance Equation

Let E be the total energy of a body, consisting of its kinetic energy K and the remaining portion I, which is called the internal energy, so that $E = K + I$. Let W be the time rate at which work is being done on the body, and Q the rate at which nonmechanical energy is entering the body. Then, the balance principle for energy, which is an expression for the first law of thermodynamics, states that

$$\dot{E} = \dot{K} + \dot{I} = W + Q \tag{2.8.1}$$

The total kinetic energy K is the sum of the kinetic energy of translation, $K_t = \int_B \frac{1}{2}\rho v_r v_r \, dV$, and the kinetic energy of spin, $K_s = \int_B \rho k_s \, dV$, where k_s is the density of spin kinetic energy per unit mass, so that $K = \int_B \rho(\frac{1}{2} v_r v_r + k_s) \, dV$. The internal energy I is also best described in terms of an internal energy density ϵ, such that $I = \int_B \rho\epsilon \, dV$. Thus, if e is the total energy per unit mass, then $E = \int_B \rho e \, dV = \int_B \rho(\epsilon + k_s + \frac{1}{2} v_r v_r) \, dV$, which implies that

$$e = \epsilon + \frac{1}{2} v_r v_r + k_s \tag{2.8.2}$$

so that

$$\dot{E} = \int_B \rho \dot{e} \, dV \quad , \quad \dot{e} = \dot{\epsilon} + v_r a_r + \dot{k}_s \tag{2.8.3}$$

Note that the term K_s is not considered in classical nonpolar theories.

The rate of working W can be divided into two parts. The contact rate W_c, due to the actions of the surface stresses t_i and m_i, is given by $W_c = \int_{\partial B} t_r v_r \, dA + \int_{\partial B} m_r \omega_r \, dA$, which on use of $t_r = t_{sr} n_s$, $m_r = m_{sr} n_s$ and the application of generalized Gauss' theorem becomes $W_c = \int_B (t_{sr} v_r)_{,s} \, dV + \int_B (m_{sr} \omega_r)_{,s} \, dV$. The body rate W_b, due to the body force and the body moment, is given by $W_b = \int_B \rho f_r v_r \, dV + \int_B \rho l_r \omega_r \, dV$. Thus

$$W = W_c + W_b = \int_B [v_r (t_{sr,s} + \rho f_r) + \omega_r (m_{sr,s} + \rho l_r) + t_{rs} v_{s,r} + m_{rs} \omega_{s,r}] \, dV \tag{2.8.4}$$

The rate at which energy is entering the body can, again, be described in terms of the contact energy influx rate Q_c on ∂B and the body influx rate Q_b, the rate at which energy is being generated inside B. The contact influx rate is best described in terms of the energy influx rate vector $\mathbf{q}$, such that $Q_c = -\int_{\partial B} q_r n_r \, dA = -\int_B q_{r,r} \, dV$. If h is the rate at which energy is being generated in B, per unit mass, then $Q_b = \int_B \rho h \, dV$, so that

$$Q = Q_c + Q_b = -\int_B q_{r,r} \, dV + \int_B \rho h \, dV \tag{2.8.5}$$

A substitution from Eqs. (2.8.3), (2.8.4) and (2.8.5) in Eq. (2.8.1) then results in

$$\int_B \rho (\dot{\epsilon} + v_r a_r + \dot{k}_s) \, dV = \int_B [v_r (t_{sr,s} + \rho f_r) + \omega_r (m_{sr,s} + \rho l_r) + t_{rs} v_{s,r} + m_{rs} \omega_{s,r} - q_{r,r} + \rho h] \, dV$$

and, since this must hold for all parts B of B_o,

$$\rho (\dot{\epsilon} + v_r a_r + \dot{k}_s) = v_r (t_{sr,s} + \rho f_r) + \omega_r (m_{sr,s} + \rho l_r) + t_{rs} v_{s,r} + m_{rs} \omega_{s,r} - q_{r,r} + \rho h \tag{2.8.6}$$

This equation can be simplified by using Cauchy's first and second laws of motion. The equation can be written as

$$\rho (\dot{\epsilon} + \dot{k}_s) + v_r (\rho a_r - t_{sr,s} - \rho f_r) = \omega_r (m_{sr,s} + \rho l_r) + t_{rs} v_{s,r} + m_{rs} \omega_{s,r} - q_{r,r} + \rho h$$

By virtue of Eq. (2.6.4), the second term on the left-hand side is identically zero. Using Eq. (2.6.5), the first term on the right-hand side becomes $\omega_r(\rho\dot{\sigma}_r - e_{rst}t_{st}) = \rho\omega_r\dot{\sigma}_r - t_{st}w_{st}$. When combined with $t_{rs}v_{s,r} = t_{rs}(d_{rs} + w_{rs})$, this term gives $\omega_r(m_{sr,s} + \rho l_r) + t_{rs}v_{s,r} = \rho\omega_r\dot{\sigma}_r + t_{rs}d_{rs}$, and, since $m_{rs}\omega_{s,r} = m_{rs}k_{rs}$, the energy equation becomes

$$\rho(\dot{\epsilon} + \dot{k}_s) = \rho\omega_r\dot{\sigma}_r + t_{rs}d_{rs} + m_{rs}k_{rs} - q_{r,r} + \rho h \tag{2.8.7}$$

Finally, since d_{rs} is symmetric and $k_{rr} = 0$,

$$\rho(\dot{\epsilon} + \dot{k}_s) = \rho\omega_r\dot{\sigma}_r + t_{(rs)}d_{rs} + m_{rs}^D k_{rs} - q_{r,r} + \rho h \tag{2.8.8}$$

where $t_{(rs)} = \frac{1}{2}(t_{rs} + t_{sr})$ is the symmetric part of $\mathbf{T}$ and $m_{rs}^D = m_{rs} - \frac{1}{3}m_{kk}\delta_{rs}$ is the deviatoric part of $\mathbf{M}$.

This result is a consequence of the energy balance equation and the first and second laws of Cauchy. An application of the energy equation alone results in Eqs. (2.8.6).

Note that the only kinematic measures that appear in the energy balance equation are the rate of deformation tensor $\mathbf{D}$ and the curvature-twist rate tensor $\mathbf{K}$. Furthermore, only the symmetric part of $\mathbf{T}$ and the deviatoric part of $\mathbf{M}$ have a direct effect on the energy equation. In addition, other kinematic measures, which describe micromotion in theories with microstructure, can come in through the k_s term.

2.9 Entropy Inequality

In addition to the internal energy per unit mass, ϵ, there are two important scalar quantities associated with thermal interactions, namely the temperature θ and the entropy per unit mass η. The total entropy of B is then given by

$$\mathrm{H} = \int_B \rho\eta \, dV \quad or \quad \dot{\mathrm{H}} = \int_B \rho\dot{\eta} \, dV \tag{2.9.1}$$

The second law of thermodynamics will be used in the form of the Clausius-Duhem inequality

$$\dot{\mathrm{H}} \geqslant -\int_{\partial B} \frac{1}{\theta} q_r n_r \, dA + \int_B \frac{1}{\theta}\rho h \, dV \tag{2.9.2}$$

where, as before, q_r is the energy influx rate vector. Here q_r/θ can be looked upon as the rate of influx of entropy across ∂B and h/θ as the entropy source. The entropy inequality thus states that the rate of increase of the entropy of a body B is always greater than, or equal to, the rate of influx of entropy across ∂B plus the rate at which entropy enters throughout the volume.

The Clausius-Duhem inequality may be converted into an equality by defining a rate of entropy production Γ through

$$\dot{\mathrm{H}} = = \int_{\partial B} \frac{1}{\theta} q_r n_r \, dA + \int_B \frac{1}{\theta} \rho h \, dV + \Gamma \tag{2.9.3}$$

Γ may be written as

$$\Gamma = \int_B \rho \gamma \, dV \tag{2.9.4}$$

where γ is the rate of entropy production per unit mass.

Substituting in Eq. (2.9.3) from Eqs. (2.9.1) and (2.9.4), and using generalized Gauss' theorem, the second law may be expressed as

$$\int_B \rho \dot{\eta} \, dV = \int_B [- (q_r/\theta)_{,r} + \frac{1}{\theta} \rho h + \rho \gamma] \, dV$$

and, since this must hold for all B,

$$\dot{\eta} = - \frac{1}{\rho} (q_r/\theta)_{,r} + \frac{h}{\theta} + \gamma \tag{2.9.5}$$

The Clausius-Duhem inequality implies that $\Gamma \geqslant 0$, so that $\gamma \geqslant 0$ follows from Eq. (2.9.4). Thus

$$\Gamma \geqslant 0 \quad , \quad \gamma \geqslant 0 \tag{2.9.6}$$

Using the energy balance equality, given by Eq. (2.8.7), the source terms containing h may be eliminated, resulting in

$$\theta \gamma = \theta \dot{\eta} + \frac{\theta}{\rho} (q_r/\theta)_{,r} - (\dot{\epsilon} + \dot{k}_s) + \omega_{,r} \dot{\sigma}_r + \frac{1}{\rho} t_{(rs)} d_{rs} + \frac{1}{\rho} m_{rs}^D k_{rs} - \frac{1}{\rho} q_{r,r}$$

which simplifies to

$$\theta \gamma = \theta \dot{\eta} - \dot{\epsilon} - \dot{k}_s + \omega_{,r} \dot{\sigma}_r + \frac{1}{\rho} t_{(rs)} d_{rs} + \frac{1}{\rho} m_{rs}^D k_{rs} - \frac{q_r \theta_{,r}}{\rho \theta} \tag{2.9.7}$$

The constraint imposed by this equation is that $\gamma \geqslant 0$ for all bodies and motions.

If the free energy $\psi = \epsilon - \theta \eta$ is used, then this equation becomes

$$\theta \gamma = -\dot{\psi} - \dot{k}_s + \omega_{,r} \dot{\sigma}_r + \frac{1}{\rho} t_{(rs)} d_{rs} + \frac{1}{\rho} m_{rs}^D k_{rs} - \eta \dot{\theta} - \frac{q_r \theta_{,r}}{\rho \theta} \tag{2.9.8}$$

The internal energy ϵ may be assumed to be a function of η and N state variables ν_α, $\alpha = 1, 2, \cdots N$, that is

$$\epsilon = \epsilon(\eta, \nu_\alpha, \mathbf{x}) \tag{2.9.9}$$

where $\eta = \eta(\mathbf{x}, t)$ and $\nu_\alpha = \nu_\alpha(\mathbf{x}, t)$. The absolute temperature θ and the thermodynamic tensions τ_α are defined by

$$\theta = \frac{\partial \epsilon}{\partial \eta}\Big|_{\nu_\alpha}, \quad \tau_\alpha = \frac{\partial \epsilon}{\partial \nu_\alpha}\Big|_{\eta, \nu_\beta\ (\beta \neq \alpha)} \tag{2.9.10}$$

so that from Eq. (2.9.9)

$$\dot{\epsilon} = \theta \dot{\eta} + \tau_\alpha \dot{\nu}_\alpha \tag{2.9.11}$$

where $\tau_\alpha \dot{\nu}_\alpha = \sum_{r=1}^{N} \tau_r \dot{\nu}_r$. Finally, from Eq. (2.9.7) γ is given by

$$\rho\theta\gamma = t_{(rs)} d_{rs} + m_{rs}^D k_{rs} + \rho\omega_r \dot{\sigma}_r - \rho \dot{k}_s - q_r (\ln \theta)_{,r} - \rho \tau_\alpha \dot{\nu}_\alpha \tag{2.9.12}$$

One application of the second law is in establishing the signs of the coefficients in constitutive equations, which is accomplished by requiring that $\gamma \geqslant 0$, that is that $\rho\theta\gamma \geqslant 0$, for all possible motions.

2.10 Concluding Remarks

In this chapter, the main field equations governing the motion of continuous media have been established for the polar case in which internal spin, couple stresses, and body moments are present. The first major departure from classical nonpolar theories is that the stress tensor is no longer symmetric. Furthermore, in addition to the nine components of the stress tensor, nine components of the couple stress tensor are also present. As a result the number of unknowns, describing mechanical interactions, increases from six to eighteen.

The energy and entropy equations discussed in Sections (2.8) and (2.9), respectively, show that in addition to couple stresses polar theories require a new tensor **K**, which is not considered in nonpolar theories. Note that **D** and **K** are associated with **T** and **M**, respectively, and there is no cross coupling of terms.

Once the equations of motion have been established, constitutive equations, which relate the mechanical interactions to the motion, have to be considered in order to complete the system of equations.

2.11 References

1. Stokes, V. K. (1972). On the Analysis of Asymmetric Stress, *J. Appl. Mech.* **39**, 1133-1136.

2. Stokes, V. K. (1980). On the Proper Vectors of Real Third Order Matrices, *J. Math. Phys. (N. Y.)* **21**, 2842-2843.

Chapter 3

Couple Stresses in Fluids

3.1 Introduction

Following the scheme described in the Introduction, the possible effects of couple stresses will first be considered in isolation by assuming that the fluid has no microstructure at the kinematic level, so that the kinematics of motion is fully determined by the velocity field.

This chapter therefore considers the study of a simple theory of couple stresses in fluids in which the intrinsic angular momentum $\boldsymbol{\sigma}$ and the kinetic energy of spin density k_s are not taken into account. Thus, whereas couple stresses will be considered, the fluid will not have any microstructure.

The theory discussed in this chapter is based on a heuristic approach, but is a special case of the more general theories that are now available. This is the simplest theory that shows all the important features and effects of couple stresses and results in equations that are similar to the Navier-Stokes equations, thereby facilitating a comparison with the results for the classical nonpolar case. The main effect of couple stresses will be to introduce a length-dependent effect that is not present in the classical nonpolar theories.

A list of equations that are of interest is useful. Once $\boldsymbol{\sigma}$ and k_s are taken to be zero, the various field equations, governing the motion of a fluid will be:

the continuity equation

$$\dot{\rho} + \rho v_{r,r} = 0 \tag{3.1.1}$$

Cauchy's first law of motion, given in Eq. (2.6.4),

$$\rho a_i = t_{ri,r} + \rho f_i \tag{3.1.2}$$

Cauchy's second law of motion, given in Eq. (2.6.5),

$$m_{ri,r} + \rho l_i + e_{irs} t_{rs} = 0 \tag{3.1.3}$$

the energy equation, given in Eq. (2.8.7),

$$\rho \dot{\epsilon} = t_{(rs)} d_{rs} + m_{rs}^D k_{rs} - q_{r,r} + \rho h \tag{3.1.4}$$

and the entropy inequality, given in Eq. (2.9.12),

$$\left.\begin{aligned}&\gamma \geqslant 0\\&\rho\theta\gamma = t_{(rs)}d_{rs} + m_{rs}^{D}k_{rs} - q_r(\ln\theta)_{,r}\end{aligned}\right\} \tag{3.1.5}$$

where ϵ has been assumed to be a function of η only, so that the ν_α's are absent.

If a purely mechanical theory is considered, the unknowns are twenty-two in number, ρ, v_r, t_{rs} and m_{rs}. So far only a total of seven equations, given in Eqs. (3.1.1) to (3.1.3) are available in these variables. This system must, of course, be closed by choosing an appropriate set of constitutive equations.

3.2 Constitutive Equations

The correct procedure for obtaining constitutive equations would be to assume that **T** and **M** are functions of $v_{i,j}$, $v_{i,jk}$ and so on, and to then use the principle of material objectivity to determine the correct kinematic measures and the allowable forms of the dependence of **T** and **M** on these measures. In a less general approach, a start could be made by choosing objective measures based on $v_{i,j}$ and $v_{i,jk}$. On the basis of the results of Section (1.7), two such appropriate measures are the objective tensors **D** and **K**. Equations (3.1.4) and (3.1.5) show that these are the only two kinematic variables that appear in the energy balance and in the entropy inequality. Furthermore, **D** occurs in combination with $\mathbf{T}^S$ and **K** in combination with $\mathbf{M}^D$, there being no cross coupling of terms. Heuristically, this would seem to indicate constitutive equations of the form $\mathbf{T} = \mathbf{f}(\mathbf{D})$ and $\mathbf{M} = \mathbf{g}(\mathbf{K})$.

In the theory of perfectly elastic solids, the principle of entropy balance results in an equation that is useful in discussing the form of constitutive equations. However, because of dissipation, this is not possible for fluids, since the balance of entropy is given by an inequality. For the elastic case, Mindlin and Tiersten [Ref. 1] obtained the constitutive equations for a linear perfectly elastic solid in the form

$$t_{ij}^S = t_{(ij)} = \lambda(\epsilon_{rr})\delta_{ij} + 2\mu\epsilon_{ij} \tag{3.2.1}$$

$$m_{ij}^D = m_{ij} - \frac{1}{3}m_{rr}\delta_{ij} = 4\eta\kappa_{ij} + 4\eta'\kappa_{ji} \tag{3.2.2}$$

where ϵ_{ij} is the small-strain tensor and κ_{ij} the gradient of the small-rotation vector, called the curvature-twist tensor.

In the classical nonpolar case, the constitutive equations for an elastic solid and a Newtonian fluid are given, respectively, by the equations

$$t_{ij} = \lambda(\epsilon_{rr})\delta_{ij} + 2\mu\epsilon_{ij}$$

$$t_{ij} = -p\delta_{ij} + \lambda(d_{rr})\delta_{ij} + 2\mu d_{ij}$$

Thus, there is a formal similarity between the two cases. Except for the pressure term, the stress tensor for a fluid can be obtained from the corresponding one for an elastic solid by replacing the strains ϵ_{ij} by the rate of strain d_{ij}.

Keeping in mind the fact that the objective tensors d_{ij} and k_{ij} are the only kinematic variables that appear in the energy equation, and also considering the results for linearly elastic solids given in Eqs. (3.2.1) and (3.2.2), the constitutive equations for a polar fluid will be assumed to have the form

$$t_{ij}^S = -p\delta_{ij} + \lambda d_{rr}\delta_{ij} + 2\mu d_{ij} \tag{3.2.3}$$

$$m_{ij}^D = 4\eta k_{ij} + 4\eta' k_{ji} \tag{3.2.4}$$

where λ, μ, η and η' are material constants.

Note that the reasons leading to the choice of these equations are purely heuristic. As mentioned earlier, this theory is known to be a special case of the more general theories that are now available and, as such, the predictions based on this simple theory are of interest.

Only the symmetric part of $\mathbf{T}$ and the deviatoric part of $\mathbf{M}$ are determined by the constitutive equations given in Eqs. (3.2.3) and (3.2.4). The trace of $\mathbf{M}$, $\mathrm{tr}\mathbf{M} = m_{rr} = m$ and the skew-symmetric part $\mathbf{T}^A$ of $\mathbf{T}$ are not determined by the constitutive equations. However, $\mathbf{T}^A$ can be determined from the equations of motion, once m is known. Thus, the only quantity not determined will be $m = m_{rr}$, which has to be determined by the boundary conditions. A similar situation arises in the case of an incompressible fluid, wherein the constitutive equations determine the stress to within an arbitrary hydrostatic term only.

The dimensions of the material constants λ and μ, which do not have the same significance as in the nonpolar case, are those of viscosity, namely, M/LT, whereas those of η and η' are those of momentum, namely ML/T. The ratio η/μ therefore has the dimensions of length square, and this material constant will be denoted by

$$l = (\eta/\mu)^{1/2} \tag{3.2.5}$$

For the present theory, since σ_r and ν_α are zero, and since only isothermal flows are being considered, Eq. (2.9.12) reduces to $\rho\theta\gamma = t_{(rs)}d_{rs} + m_{rs}^D k_{rs}$, which, upon using the constitutive equations given in Eqs. (3.2.3) and (3.2.4), becomes

$$\rho\theta\gamma = -pd_{rr} + \lambda d_{rr}d_{ss} + 2\mu d_{rs}d_{rs} + 4\eta k_{rs}k_{rs} + 4\eta' k_{rs}k_{sr} \tag{3.2.6}$$

Since d_{ij} and k_{ij} are independent of each other, a satisfaction of the Clausius-Duhem inequality, $\gamma \geqslant 0$, requires that

$$-pd_{rr} \geqslant 0 \tag{3.2.7}$$

$$\lambda\, d_{rr}\, d_{ss} + 2\mu\, d_{rs}\, d_{rs} \geqslant 0 \tag{3.2.8}$$

$$4\eta\, k_{rs}\, k_{rs} + 4\eta' k_{rs}\, k_{sr} \geqslant 0 \tag{3.2.9}$$

for all possible d_{rs} and k_{rs}. Equation (3.2.8) can be shown to require that

$$\mu \geqslant 0 \quad , \quad (3\lambda + 2\mu) \geqslant 0 \tag{3.2.10}$$

Similarly Eq. (3.2.9) can be shown to lead to

$$\eta + \eta' \geqslant 0 \quad , \quad \eta - \eta' \geqslant 0 \tag{3.2.11}$$

which imply

$$\eta \geqslant 0 \quad , \quad \eta \geqslant \eta' \tag{3.2.12}$$

Now that the constitutive equations have been chosen, the next step is to establish the equations of motion in terms of the velocity field. In the model that has been chosen, the couple stresses **M** depend on the vorticity gradient tensor **K**. The couple stresses may therefore be expected to be large in those nonpolar flows in which the vorticity gradients are large. The unknowns are now 39 in number: ρ, v_i, ω_i, d_{ij}, k_{ij}, t_{ij} and m_{ij}. Relating them are the 38 equations: the continuity equation, given by Eq. (3.1.1), gives one equation; Cauchy's first and second laws of motion, given by Eqs. (3.1.2) and (3.1.3), give six equations; the constitutive equations, given by Eqs. (3.2.8) and (3.2.4), together give 6 + 8 = 14 equations; the remaining 17 equations are given by the 6 kinematic relations $d_{ij} = \frac{1}{2}(v_{j,i} + v_{i,j})$, the three equations $\omega_i = \frac{1}{2} e_{ijk}\, v_{k,j}$, and the eight relations $k_{ij} = \omega_{j,i}$. There are thus only 38 equations in the 39 unknowns. The degree of indeterminacy is explained by the fact that $m = m_{rr}$ is to be determined by the boundary conditions.

3.3 Equations of Motion

The equations of motion are obtained, in terms of the velocity, by using the constitutive equations to express $t_{ri,r}$, in Cauchy's first law of motion, in terms of the velocity. Now in $t_{ri,r} = t^S_{ri,r} + t^A_{ri,r}$, $t^S_{ri,r}$ can be determined directly from Eq. (3.2.3) to give $t^S_{ri,r} = -p_{,r}\delta_{ri} + \lambda\, d_{ss,r}\delta_{ri} + 2\mu\, d_{ri,r}$, which simplifies to

$$t^S_{ri,r} = -p_{,i} + (\lambda + \mu)\, v_{r,ri} + \mu\, v_{i,rr} \tag{3.3.1}$$

From Cauchy's second law

$$e_{ris} m_{ts,t} + \rho e_{ris} l_s + e_{ris} e_{spq} t_{pq} = 0$$

which, on simplification, gives

$$t_{ri}^A = -\frac{1}{2} e_{ris}(m_{ts,t} + \rho l_s) \tag{3.3.2}$$

With $m = m_{rr}$ Eq. (3.2.4) gives $m_{ts} = m\delta_{ts} + 4\eta k_{ts} + 4\eta' k_{st}$, which, together with $k_{rs} = \omega_{s,r}$, reduces Eq. (3.3.2) to

$$t_{ri}^A = -\frac{1}{2}[e_{ris} m_{,s} + 4\eta e_{ris} \omega_{s,tt}] + \rho e_{ris} l_s \tag{3.3.3}$$

as $\omega_{t,t} = 0$. Use of $e_{ris}\omega_s = w_{ri}$ transforms the above expression to

$$t_{ri}^A = -2\eta w_{ri,tt} + \frac{1}{2} e_{irs}(m_{,s} + \rho l_s) \tag{3.3.4}$$

so that $t_{ri,r}^A = -2\eta w_{ri,rtt} + \frac{1}{2} e_{irs}(\rho l_s)_{,r}$ since the term $e_{ris} m_{,sr}$ must be zero. Finally, since $2w_{ri} = v_{i,r} - v_{r,i}$,

$$t_{ri,r}^A = -\eta v_{i,rrtt} + \eta v_{r,ritt} + \frac{1}{2} e_{irs}(\rho l_s)_{,r} \tag{3.3.5}$$

Substitution from Eqs. (3.3.1) and (3.3.4) in Cauchy's first law of motion, $\rho a_i = t_{ri,r} + \rho f_i$, then results in the equations of motion

$$\rho a_i = -p_{,i} + (\lambda + \mu)(v_{r,r})_{,i} + \mu v_{i,rr} - \eta v_{i,rrss}$$

$$+ \eta(v_{r,r})_{,iss} + \frac{1}{2} e_{irs}(\rho l_s)_{,r} + \rho f_i \tag{3.3.6}$$

which, in the notation of Gibbs, can be written as

$$\rho \mathbf{a} = -\nabla p + (\lambda + \mu)\nabla\nabla\cdot\mathbf{v} + \eta\nabla^2\nabla\nabla\cdot\mathbf{v} + \mu\nabla^2\mathbf{v} - \eta\nabla^4\mathbf{v}$$

$$+ \rho\mathbf{f} + \frac{1}{2}\nabla\times(\rho\mathbf{l}) \tag{3.3.7}$$

where $\mathbf{a} = \partial\mathbf{v}/\partial t + v\cdot\nabla\mathbf{v}$.

For incompressible fluids $\nabla\cdot\mathbf{v} = 0$. Then, if the body force $\mathbf{f}$ and the body moment $\mathbf{l}$ are absent, the equations of motion reduce to

$$\rho\mathbf{a} = -\nabla p + \mu\nabla^2\mathbf{v} - \eta\nabla^4\mathbf{v} \tag{3.3.8}$$

The last term in this equation gives the effect of couple stresses. Thus, for the effects of couple stresses to be present, $v_{i,rrss}$ must be nonzero.

From Eqs. (3.2.3) and (3.3.4)

$$t_{ij} = t_{ij}^S + t_{ij}^A = -p\delta_{ij} + \mu(v_{j,i} + v_{i,j}) - \eta v_{j,irr} + \eta v_{i,jrr}$$
$$- \frac{1}{2} e_{ijr}(m_{,r} + \rho l_r) \tag{3.3.9}$$

This equation shows one of the weak points of the constitutive equations (3.2.3) and (3.2.4) in that the stress depends on the body moment.

3.4 Boundary Conditions

For a complete description of the problem, six boundary conditions are needed for Eq. (3.3.8). Three conditions are provided by the usual assumption of no-slip at the boundary, that is, that the velocity of the fluid, relative to the surface, be zero at all solid boundaries.

In the absence of additional information, the remaining three conditions may be chosen on a heuristic basis. One possibility is to assume that the effect of boundary walls is such that, in addition to there being no slip for the velocity, fluid elements are not permitted to rotate relative to the boundary, thereby implying that the vorticity at the boundary be equal to the rate of rotation of the boundary. Another possibility is to assume that the couple stresses vanish at the walls. The two sets will be referred to as boundary conditions A and B.

Boundary Conditions A: Couple stresses zero at the boundary. This is equivalent to assuming that mechanical interactions at the wall are equipollent to a force distribution only.

Boundary Conditions B: Vorticity at the boundary equal to the rate of rotation of the boundary. As mentioned earlier, these conditions are equivalent to assuming that the effect of the wall is not only to prevent any relative velocity, but also to prevent relative rotations of the fluid elements at the wall. For some time it was thought that if the vorticity were equated to the angular velocity of the boundary at the boundary, then the vorticity would remain so, even in the limit when the couple stresses tend to zero. Thus the solution would not approach that of the Navier-Stokes equations as $\eta \to 0$. However, this does not occur, and the condition $\boldsymbol{\omega} = \boldsymbol{\omega}_B$ at the boundary is consistent in the sense that as $\eta \to 0$, the solution tends to that of the Navier-Stokes equations.

The boundary conditions A essentially restrict the gradients of $\boldsymbol{\omega}$ at the wall while the conditions B impose a restriction on $\boldsymbol{\omega}$ itself. Mixed boundary conditions of the type $m_{ri}n_r + \alpha\omega_i = 0$ on the boundary are also possible. However, only the conditions A and B will be used in this chapter.

Now that the boundary conditions have been defined, solutions illustrating the effects of couple stresses will be obtained. For the remainder of this chapter the fluid will be assumed to be incompressible, so that the equations of motion will be given by Eq. (3.3.6).

3.5 Steady Flow Between Parallel Plates

For the steady one-dimensional flow of an incompressible fluid between two parallel plates, the velocity field is of the form

$$v_1 = u(x_2) = u(y) \quad , \quad v_2 = 0 \quad , \quad v_3 = 0 \tag{3.5.1}$$

where the only nonzero component of the velocity is in the x_1 direction. This field satisfies the continuity equation identically. The equations of motion, given by Eq. (3.3.8), then reduce to

$$\frac{d^4u}{dy^4} - \frac{1}{l^2}\frac{d^2u}{dy^2} = -\frac{1}{\eta}\frac{dp}{dx} \quad , \quad p = p(x) \tag{3.5.2}$$

the general solution of which is given by

$$u = A_0 + A_1 y + \frac{1}{2\mu}\frac{dp}{dx}y^2 + B_1\cosh(y/l) + B_2\sinh(y/l) \tag{3.5.3}$$

The only nonzero components of $\boldsymbol{\omega}$, $\mathbf{K}$ and $\mathbf{M}$ are then given by

$$\omega_z = -\frac{1}{2}\frac{du}{dy} = -\frac{1}{2}\left[A_1 + \frac{1}{\mu}\frac{dp}{dx}y + \frac{B_1}{l}\sinh(y/l) + \frac{B_2}{l}\cosh(y/l)\right] \tag{3.5.4}$$

$$k_{yz} = \frac{d\omega_z}{dy} = -\frac{1}{2}\left[\frac{1}{\mu}\frac{dp}{dx} + \frac{B_1}{l^2}\cosh(y/l) + \frac{B_2}{l^2}\sinh(y/l)\right] \tag{3.5.5}$$

$$m_{yz} = 4\eta k_{yz} = -2\left[l^2\frac{dp}{dx} + B_1\cosh(y/l) + B_2\sinh(y/l)\right] \tag{3.5.6}$$

From the constitutive equations given in Eq. (3.2.3), $t^S_{yx} = 2\mu d_{yx} = \mu\, du/dy$. Also, from Eq. (3.3.4) $t^A_{yx} = 2\eta d^2\omega_z/dy^2 = 2\eta\, dk_{yz}/dy$, so that

$$t^S_{yx} = \mu\left[A_1 + \frac{1}{\mu}\frac{dp}{dx}y + \frac{B_1}{l}\sinh(y/l) + \frac{B_2}{l}\cosh(y/l)\right]$$

$$t^A_{yx} = -\mu\left[\frac{B_1}{l}\sinh(y/l) + \frac{B_2}{l}\cosh(y/l)\right]$$

Then

$$t_{yx} = t^S_{yx} + t^A_{yx} = \mu A_1 + \frac{dp}{dx}y \tag{3.5.7}$$

Thus, for the particular flows under consideration, the wall shear — the value of t_{yx} at the walls, can depend on couple stresses only through the constant A_1.

Note that the material constant η' will not affect any of the properties of the flow except the couple stress component m_{zy}. Thus, these flows do not provide a method for measuring η'.

1. Couette Flow with dp/dx = 0

Consider the particular case in which the top plate moves with a constant velocity V in the positive x direction, the bottom plate being fixed. Let the distance between the plate be h.

Boundary Conditions A:

For boundary conditions A, the general solution of the equations of motion, given by Eq. (3.5.3), must satisfy

$$u(0) = 0 \quad , \quad u(h) = V$$

$$m_{yz}(0) = 0 \quad , \quad m_{yz}(h) = 0 \tag{3.5.8}$$

The conditions on m_{zy} imply that $k_{yz}(0) = 0$ and $k_{zy}(h) = 0$. Under these conditions, the general solution reduces to

$$u = \frac{V}{h} y \quad , \quad \omega_z = -\frac{1}{2}\frac{V}{h} \quad , \quad k_{yz} = 0 \tag{3.5.9}$$

The couple stresses are identically zero. The wall shear is given by $\tau_w = t_{yx}(h) = \mu \; du/dy|_{y=h} = \mu V/h$, so that

$$\mu = \frac{h\tau_w}{V} \tag{3.5.10}$$

Thus, in this particular flow, the couple stresses are identically zero. This flow defines μ experimentally.

Boundary Conditions B:

The boundary conditions for this case are

$$u(0) = 0 \quad , \quad u(h) = V$$

$$\omega_z(0) = 0 \quad , \quad \omega_z(h) = 0 \tag{3.5.11}$$

for which the solution is given by

$$U(\xi) = u(\xi)/V = \frac{1}{D(a)} [\xi a \sinh a - \{\cosh a + \cosh a\xi$$

$$- \cosh a(1-\xi) - 1\}] \quad (3.5.12)$$

where $\xi = y/h$, $a = h/l$, and $D(a) = a \sinh a - 2(\cosh a - 1)$. The vorticity and couple stress distributions are then given by

$$\Omega(\xi) = \omega_z/(-V/2h) = \frac{a}{D(a)}[\sinh a - \sinh a\xi - \sinh a(1-\xi)] \quad (3.5.13)$$

$$M(\xi) = m_{yz}(\xi)/2\mu V = \frac{1}{D(a)}[\cosh a\xi - \cosh a(1-\xi)] \quad (3.5.14)$$

From Eq. (3.5.7) the wall shear, $\tau_w = t_{yx}(h)$, for this geometry is given by

$$\frac{\tau_w}{\mu V/h} = \frac{t_{yx}(h)}{\mu V/h} = \frac{1}{1 - \frac{2}{a}(\coth a - 1/\sinh a)} \quad (3.5.15)$$

Thus, for boundary conditions B, couple stresses do have an effect on plane Couette flow, the main feature being that a size-dependent effect is introduced, so that even after a nondimensionalization of the equations, the velocity distribution is affected by the nondimensional parameter $a = h/l$. The variations of the velocity and vorticity profiles with ξ, and a as parameter, are shown in Figs. 3.5.1 and 3.5.2, respectively. As a increases, the effects of couple stresses decrease. In fact the limit of $a \to \infty$ is given by $\lim_{a\to\infty} U(\xi) = \xi$. This may also be interpreted as an increase in the gap size

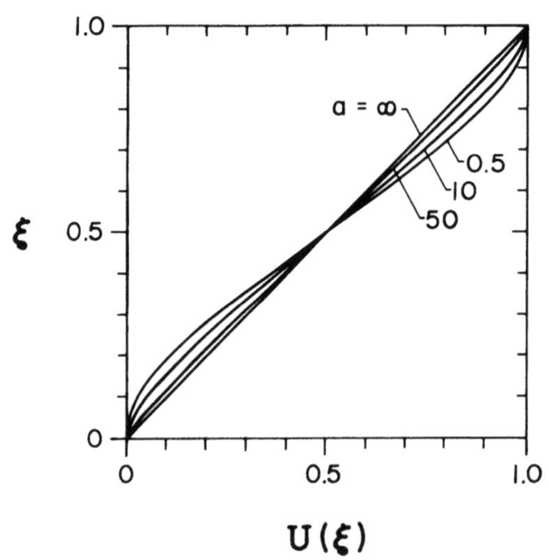

Fig. 3.5.1 Variation of the velocity $U(\xi)$ for plane Couette flow with boundary conditions B.

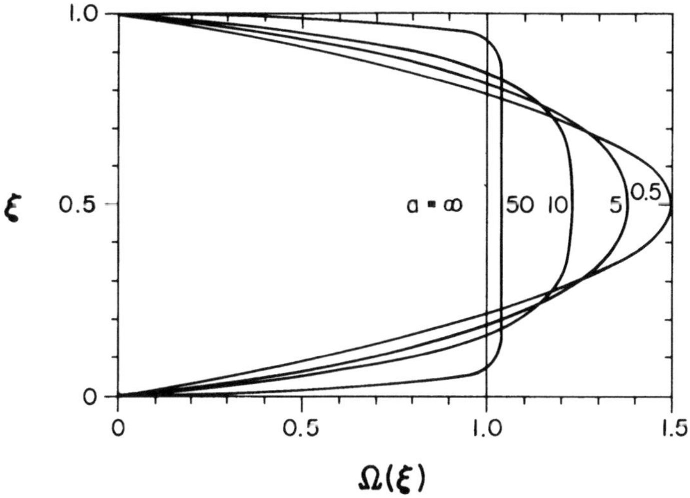

Fig. 3.5.2 Variation of the vorticity $\Omega(\xi)$ for plane Couette flow with boundary conditions B.

h. As a decreases, the effects of couple stresses increase. The limiting case of $a \to 0$ resuits in $\lim_{a\to 0} U(\xi) = \xi^2(3 - 2\xi)$, $\lim_{a \to 0} \Omega(\xi) = 6\xi(1 - \xi)$, and in the couple stresses become infinite.

Note that, as $a \to \infty$, the vorticity distribution approaches the nonpolar distribution $\Omega \equiv 1$. In fact, in the limiting case $\Omega \equiv 1$, except for the singularities $\Omega \equiv 0$ at $\xi = 0$ and $\xi = 1$ due to the imposition of the boundary conditions.

Equation (3.5.14) shows that the couple stress m_{yz} is skew-symmetric about the mid-plane $\xi = 0.5$, being zero at $\xi = 0.5$, and takes on its maximum value, m_{max}, at $\xi = 0$ and $\xi = 1$. The variation of $M_{max} = m_{max}/2\mu V = (\cosh a - 1)/D(a)$ is shown in Fig. 3.5.3. The cou-

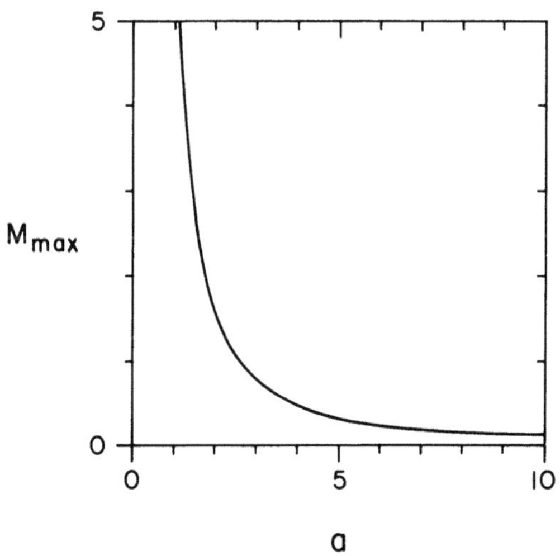

Fig. 3.5.3 Variation of the maximum couple stress M_{max} versus a.

ple stress is small for large values of a but increases rapidly as the value of a decreases. The variation of m_{yz} is more or less linear between $-m_{\max}$ at $\xi = 0$ to $m_{\max}$ at $\xi = 1$, being exactly zero at $\xi = 0.5$.

Equation (3.5.15) shows that the wall shear does not measure μ, but rather a combination of μ and η. However, as $a \to \infty$, $t_{yx}/(\mu V h) \to 1$ so that $\mu = \underset{a \to \infty}{\text{limit}} (\tau_w h/V)$. Thus, in principle, if τ_w is measured for different V, for increasing values of h, then $\mu = \underset{h \to \infty}{\text{limit}} (\tau_w h/V)$. Of course, this limit would stabilize to a constant value even for a as low as 50. Once μ has been determined, η can be obtained by solving Eq. (3.5.15).

Since the solution for boundary conditions A is the same as the nonpolar solution, there being no couple stresses, this flow, in principle, can be used for testing the validity of boundary conditions A and B.

2. Plane Poiseuille Flow

Consider the flow between two parallel plates, which are at a distance $2h$ from each other, due to a constant pressure gradient dp/dx. Let the $x = x_1$ axis coincide with the centerline, so that the plates are at $y = \pm h$.

Boundary Conditions A:

For this case the appropriate boundary conditions are

$$u(\pm h) = 0 \quad , \quad m_{yz}(\pm h) = 0 \tag{3.5.16}$$

Subject to these conditions, the general solution, given by Eqs. (3.5.3) to (3.5.6), reduces to

$$U(\xi) = \frac{u(\xi)}{\dfrac{h^2}{2\mu}\dfrac{dp}{dx}} = 1 - \xi^2 - \frac{2}{a^2}\left[1 - \frac{\cosh a\xi}{\cosh a}\right] \tag{3.5.17}$$

$$\Omega(\xi) = \frac{\omega_z(\xi)}{\dfrac{-h}{2\mu}\dfrac{dp}{dx}} = \xi - \frac{1}{a}\,\frac{\sinh a\xi}{\cosh a} \tag{3.5.18}$$

$$M(\xi) = \frac{m_{yz}}{-2h^2\dfrac{dp}{dx}} = \frac{1}{a^2}\left[1 - \frac{\cosh a\xi}{\cosh a}\right] \tag{3.5.19}$$

where $\xi = y/h$ and $a = h/l$.

The limiting values of $U(\xi)$ can be shown to be $\underset{a \to \infty}{\text{limit}}\ U(\xi) = 1 - \xi^2$ and $\underset{a \to 0}{\text{limit}}\ U(\xi) \equiv 0$. The velocity distribution, as a function of the parameter a, is shown in Fig. 3.5.4. The effect of the couple stresses is to decrease the velocity. Here again, the nondimensional velocity $U(\xi)$ depends on the

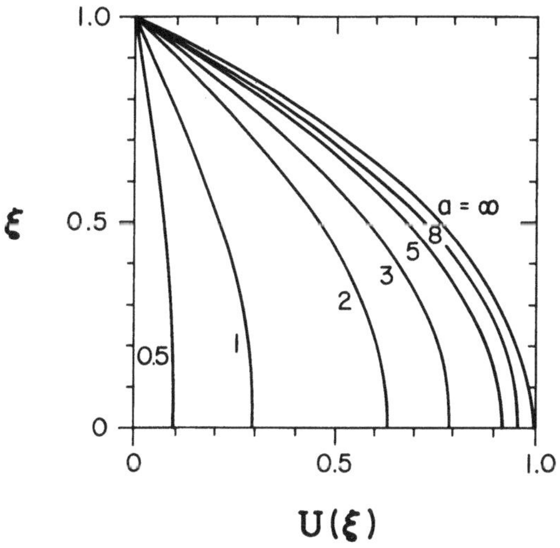

Fig. 3.5.4 Variation of the velocity $U(\xi)$ for plane Poiseuille flow with boundary conditions A. *Adapted from Ref. [3].*

value of $a = h/l$ and hence on h. This size-dependent effect is not predicted by the nonpolar theory.

Equation (3.5.18) can be used to show that $\lim_{a \to \infty} \Omega(\xi) = \xi$, the vorticity for the nonpolar case. The variation of $\Omega(\xi)$ versus ξ, with a as parameter, is shown in Fig. 3.5.5. $\Omega(\xi)$ approaches $\Omega = \xi$ as a increases.

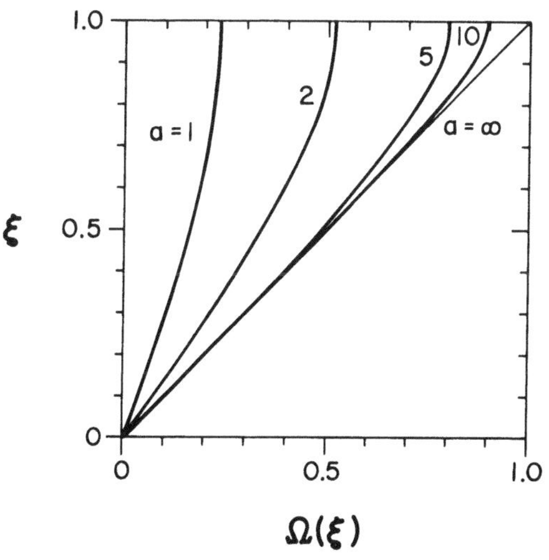

Fig. 3.5.5 Variation of the vorticity $\Omega(\xi)$ for plane Poiseuille flow with boundary conditions A.

The couple stresses are a maximum at $\xi = 0$, being zero at $\xi = \pm 1$. The variation of $M(\xi) = m_{yz}/(-2h^2 dp/dx) = (1 - \cosh a\xi/\cosh a)/a^2$ is shown in Fig. 3.5.6. The couple stresses are small for large values of a and vice versa.

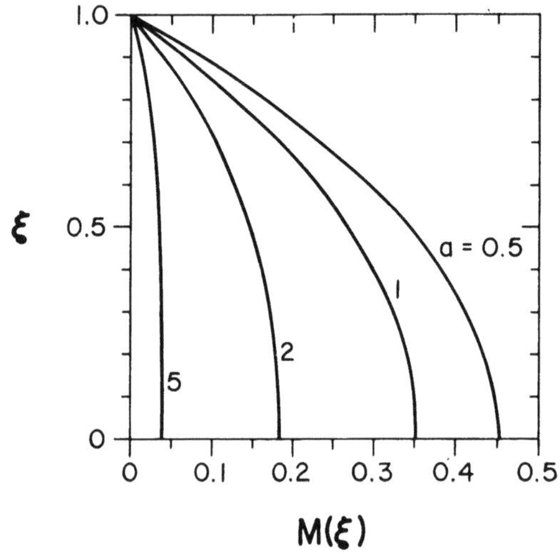

Fig. 3.5.6 Variation of the couple stress $M(\xi)$ for plane Poiseuille flow with boundary conditions A.

If Q is the volumetric through flow between the plates, per unit depth, then $Q = \int_0^{2h} u(y)\,dy$ gives

$$Q = -\frac{2}{3}\frac{h^3}{\mu}\frac{dp}{dx}\left[1 - \frac{3}{a^2}\left\{1 - \frac{1}{a}\tanh a\right\}\right] \tag{3.5.20}$$

In the absence of couple stresses, the flow would be given by

$$Q_o = -\frac{2}{3}\frac{h^3}{\mu}\frac{dp}{dx} \tag{3.5.21}$$

so that

$$\frac{Q}{Q_o} = 1 - \frac{3}{a^2}\left[1 - \frac{1}{a}\tanh a\right] \tag{3.5.22}$$

The limiting values of Q/Q_o are given by $\underset{a \to 0}{limit}\,(Q/Q_o) = 0$, $\underset{a \to \infty}{limit}\,(Q/Q_o) = 1$, $\underset{a \to 0}{limit}\left[\frac{d}{da}(Q/Q_o)\right] = 0$, and $\underset{a \to \infty}{limit}\left[\frac{d}{da}(Q/Q_o)\right] = 0$.

The variation of Q/Q_o is shown in Fig. 3.5.7.

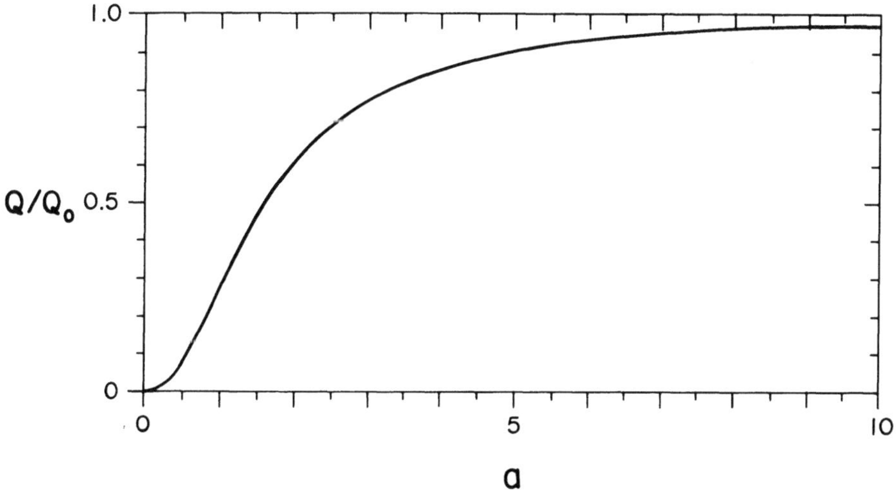

Fig. 3.5.7 Variation of Q/Q_0 versus a for plane Poiseuille flow with boundary conditions A. *Adapted from Ref. [3].*

For boundary conditions A, μ can be determined, in principle, by using experimental data for plane Couette flow. Once μ is known, so also is Q_o. If Q is measured in a plane Poiseuille flow experiment, then Eq. (3.5.22) may be used for obtaining a and hence η. Thus, in principle, η can be determined by using this flow.

Even if μ is not known from other experiments, both μ and η may be obtained from plane Poiseuille flow experiments. For, $Q/Q_o \to 1$ as $a \to \infty$, that is as $h \to \infty$, so that μ is defined by $\mu = \underset{h \to \infty}{limit} \left[-\frac{2}{3}(h^3/Q)\, dp/dx \right]$. In fact, the plot in Fig. 3.5.7 shows that even for a as low as 10 - 50, the above expression would converge to μ. Thus, in principle, μ can be determined if data is obtained for plane Poiseuille flow for different dp/dx and different h. The effects of couple stresses will be larger for smaller values of h and will decrease with the increase in h. The increase in h need only be continued until a limiting value of μ has been obtained.

The wall shear τ_w will be the value of t_{yx} at the walls. From Eqs. (3.5.7)

$$t_{yx} = \frac{dp}{dx}\, y \tag{3.5.23}$$

which is the same as the distribution for the nonpolar case. The expression for the wall shear is therefore $\tau_w = h\, dp/dx$, which is the same as in the nonpolar case. Note that even though the through flow is affected by couple stresses, the wall shear is not.

Boundary Conditions B:
The boundary conditions for this case are

$$u(\pm h) = 0 \quad , \quad \omega_z(\pm h) = 0 \tag{3.5.24}$$

and the general solution reduces to

$$U(\xi) = \frac{u(\xi)}{\dfrac{-h^2}{2\mu}\dfrac{dp}{dx}} = 1 - \xi^2 - \frac{2}{a}\left[\frac{\cosh a - \cosh a\xi}{\sinh a}\right] \tag{3.5.25}$$

$$\Omega(\xi) = \frac{\omega_z(\xi)}{\dfrac{-h}{2\mu}\dfrac{dp}{dx}} = \xi - \frac{\sinh a\xi}{\sinh a} \tag{3.5.26}$$

$$M(\xi) = \frac{m_{yz}(\xi)}{-2h^2\dfrac{dp}{dx}} = \frac{1}{a^2}\left[1 - \frac{a\cosh a\xi}{\sinh a}\right] \tag{3.5.27}$$

$$\frac{Q}{Q_o} = 1 - \frac{3}{a^2}\left[a\coth a - 1\right] \tag{3.5.28}$$

$$\tau_w = h\,\frac{dp}{dx} \tag{3.5.29}$$

In the limit as $a \to \infty$, $\sinh a\xi/\sinh a \to 0$ since $|\xi| \leqslant 1$, so that $\lim_{a\to\infty} \Omega(\xi) = \xi$, which agrees with the value for the nonpolar case. The variations of the velocity distribution with ξ, as a function of the parameter a, are shown in Fig. 3.5.8. Comparing this figure with Fig. 3.5.4 shows that

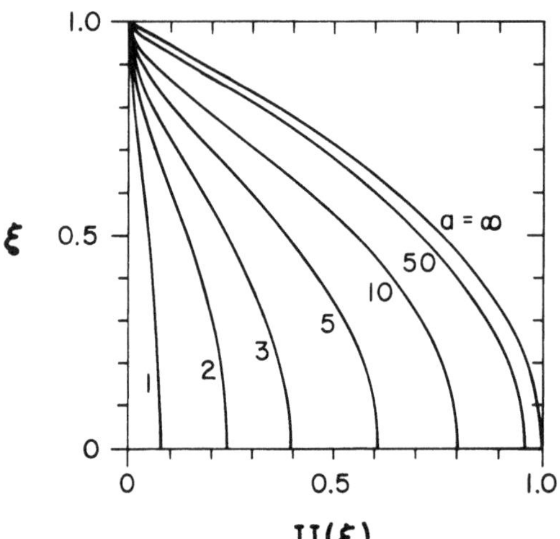

Fig. 3.5.8 Variation of the velocity $U(\xi)$ for plane Poiseuille flow with boundary conditions B.

the effects of couple stresses are more marked for boundary conditions B than for conditions A. For example, for conditions B, the effect of couple stress is noticeable even for $a = 50$, while for conditions A, the effect is smaller even for a as low as 10. The vorticity distribution is shown in Fig. 3.5.9. As a increases, Ω approaches the nonpolar value of $\Omega = \xi$ and,

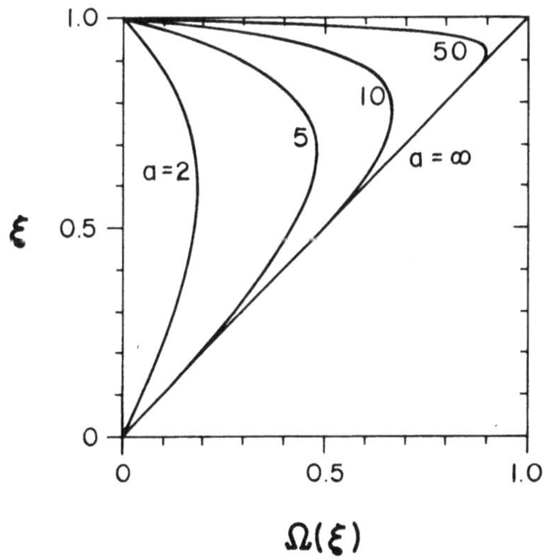

Fig. 3.5.9 Variation of the vorticity $\Omega(\xi)$ for plane Poiseuille flow with boundary conditions B.

in the limit $a \to \infty$, $\Omega = \xi$ except for the singularities at $\xi = \pm 1$ due to the imposition of the boundary conditions $\Omega(\pm 1) = 0$.

The variation of the couple stress for $a = 0.5$, 1, 2 and 5 is shown in Fig. 3.5.10. The couple stresses decrease as a increases.

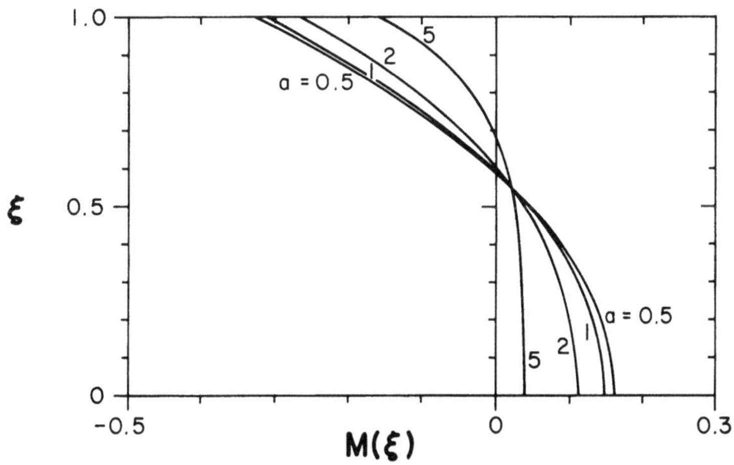

Fig. 3.5.10 Variation of the couple stress $M(\xi)$ for plane Poiseuille flow with boundary conditions B.

The variation of the expression for Q/Q_o, given in Eq. (3.5.28), has been shown in Fig. 3.5.11 by a solid line. The dotted line indicates the variation of Q/Q_o for boundary conditions A, as given in Eq. (3.5.22), and shows that the use of boundary conditions B predicts a larger effect of couple stresses than for boundary conditions A.

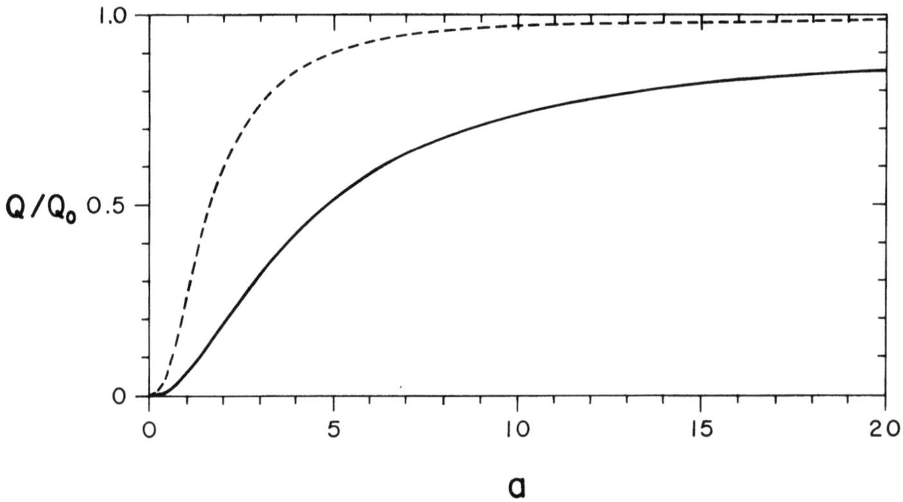

Fig. 3.5.11 Variation of Q/Q_0 versus a for plane Poiseuille flow with boundary conditions B.

Thus, the plane flows discussed in this section can, in principle, be used for determining η and μ as well as for checking for the validity of the boundary conditions. These flows are unaffected by the material constant η'. The solutions that have been presented show that the effects of couple stresses are more marked for boundary conditions B than for conditions A.

3.6 Steady Tangential Flow Between Two Coaxial Cylinders

In cylindrical polar coordinates (r,θ,z) consider steady flows of the type in which the components of the velocity are of the form

$$v_r = 0 \quad , \quad v_\theta = u(r) \quad , \quad v_z = 0 \tag{3.6.1}$$

Then, it can be shown that the equations of motion reduce to

$$\frac{\partial p}{\partial r} = \frac{u^2}{r} \quad , \quad \nabla^4 u - \frac{1}{l^2}\nabla^2 u = -\frac{1}{\eta r}\frac{\partial p}{\partial \theta} \quad , \quad \frac{\partial p}{\partial z} = 0 \tag{3.6.2}$$

where $\nabla^2 u \equiv \frac{\partial}{\partial r}\left[\frac{1}{r}\frac{\partial}{\partial r}(ru)\right]$.

The general solution of this equation is

$$u = \frac{1}{4\mu}\frac{\partial p}{\partial \theta} r(\ln r^2 - 1) + \frac{A_1}{r} + A_2 r + A_3 I_1(r/l) + A_4 K_1(r/l) \tag{3.6.3}$$

where A_1, A_2, A_3 and A_4 are arbitrary constants, and I_1 and K_1 are the modified Bessel functions of order one, of the first and second kind, respectively. The only nonzero components of $\boldsymbol{\omega}$, $\mathbf{K}$, and $\mathbf{M}$ are then given by

$$\omega_z = \frac{1}{2r}\frac{\partial}{\partial r}(ru) = \frac{1}{2\mu}\frac{\partial p}{\partial \theta}\ln r + A_2 + \frac{A_3}{2l} I_0(r/l) - \frac{A_4}{2l} K_0(r/l) \tag{3.6.4}$$

$$k_{rz} = \frac{\partial \omega_z}{\partial r} = \frac{1}{2\mu r}\frac{\partial p}{\partial \theta} + \frac{A_3}{2l^2} I_1(r/l) + \frac{A_4}{2l^2} K_1(r/l) \tag{3.6.5}$$

$$m_{rz} = 4\eta k_{rz} \quad , \quad m_{zr} = 4\eta' k_{rz} \tag{3.6.6}$$

By using the relations $t^S_{r\theta} = 2\mu d_{r\theta} = \mu r \frac{\partial}{\partial r}(u/r)$ and $t^A_{r\theta} = -\frac{1}{2r}\frac{\partial}{\partial r}(r m_{rz}) = -\frac{2\eta}{r}\frac{\partial}{\partial r}(r k_{rz})$,

$$t_{r\theta} = t^S_{r\theta} + t^A_{r\theta} = \frac{1}{2}\frac{\partial p}{\partial \theta} - 2\mu\left[\frac{A_1}{r^2} + \frac{A_3}{r} I_1(r/l) + \frac{A_4}{r} K_1(r/l)\right] \tag{3.6.7}$$

In contrast to the one-dimensional plane flows considered in the last section, the effect of couple stresses comes in explicitly in the expression for $t_{r\theta}$ because of the terms which have $I_1(r/l)$ and $K_1(r/l)$.

Here again, the material constant η' will not affect any of the properties of the flow except the couple stress m_{zr}. Thus, even these flows do not provide a method for determining η'.

1. Couette Flow with $\partial p/\partial \theta = 0$

In this case the flow is caused by one of the cylinders rotating at a constant angular velocity Ω. Let the outer cylinder rotate while the inner cylinder is kept stationary. Let the radii of the inner and outer cylinders be λR and R, respectively, where $\lambda < 1$.

Boundary Conditions A:

For this case the appropriate boundary conditions are

$$u(\lambda R) = 0 \quad , \quad m_{rz}(\lambda R) = 0$$

$$u(R) = R\lambda \quad , \quad m_{rz}(R) = 0 \tag{3.6.8}$$

Here, as in the case of plane Couette flow between two parallel plates, the couple stresses vanish throughout and the general solution reduces to

$$v_\theta = \frac{\Omega}{1-\lambda^2}[1 - \lambda^2(R/r)]r \tag{3.6.9}$$

$$\omega_z = \frac{\Omega}{1-\lambda^2} \tag{3.6.10}$$

The shear stress is given by $t_{r\theta} = 2\mu\lambda^2\Omega\,(R/r)^2/(1-\lambda^2)$, so that the wall shear at the inner cylinder is given by $\tau_w = 2\mu\Omega/(1-\lambda^2)$. If the total torque per unit height of the cylinder is T, then $T = 2\pi\lambda^2R^2\tau_w$, so that

$$\mu = \frac{1-\lambda^2}{4\pi\lambda^2R^2\Omega}\,T \tag{3.6.11}$$

This result could therefore be used for determining μ experimentally.

Boundary Conditions B:
The boundary conditions for this case are

$$\begin{aligned} u(\lambda R) &= 0 \quad , \quad \omega_z(\lambda R) = 0 \\ u(R) &= R\Omega \quad , \quad \omega_z(R) = \Omega \end{aligned} \tag{3.6.12}$$

The solution is then given by Eqs. (3.6.3) to (3.6.6), with $\partial p/\partial\theta = 0$, where the constants A_1, A_2, A_3 and A_4 are given by

$$\left.\begin{aligned} A_1 &= R\Omega\,\frac{\lambda R}{D}\left[Z_3\left\{I_1(\lambda a) - \frac{\lambda a}{2}\,I_0(\lambda a)\right\}\right. \\ &\qquad\left. - Y_3\left\{K_1(\lambda a) + \frac{\lambda a}{2}\,K_0(\lambda a)\right\}\right] \\ A_2 &= R\Omega\,\frac{a}{2RD}\left[Z_3I_0(\lambda a) + Y_3K_0(\lambda a)\right] \\ A_3 &= -\frac{R\Omega}{D}\,Z_3\,,\ A_4 = \frac{R\Omega}{D}\,Y_3 \end{aligned}\right\} \tag{3.6.13}$$

where

$$\left.\begin{aligned}
&D = Y_0 Z_2 + Y_2 Z_0 \\
&\left.\begin{aligned} Y_n &= I_n(a) - \lambda^n I_n(\lambda a) \\ Z_n &= K_n(a) - \lambda^n K_n(\lambda a) \end{aligned}\right\} n = 0, 1 \\
&Y_2 = Y_1 - (1 - \lambda^2)\,\frac{a}{2}\,I_0(\lambda a) \\
&Z_2 = Z_1 + (1 - \lambda^2)\,\frac{a}{2}\,K_0(\lambda a) \\
&Y_3 = Y_0 - \frac{2}{a}\,Y_2 \quad , \quad Z_3 = Z_0 - \frac{2}{a}\,Z_2
\end{aligned}\right\} \tag{3.6.14}$$

Then, from Eq. (3.6.7) the wall shear, $\tau_w = t_{r\theta}(\lambda R)$, at $r = \lambda R$ is given by

$$\tau_w = \frac{\mu \Omega a}{D}\,[Z_3 I_0(\lambda a) + Y_3 K_0(\lambda a)] \tag{3.6.15}$$

In this case, a measurement of the torque $T = 2\pi\lambda^2 R^2 \tau_w$ does not determine μ since τ_w is a function of l. A Couette viscometer would therefore give a combined measure for μ and η.

Thus there is a similarity between plane and cylindrical Couette flow in that both the flows are unaffected by couple stresses for boundary conditions A, while both are affected for boundary conditions B. In principle, these flows provide a method for checking the validity of the boundary conditions experimentally.

2. Poiseuille Flow Due to a Toroidal Pressure Gradient

In this case the flow is caused by a toroidal pressure gradient, $\partial p/\partial\theta$, which acts between two stationary cylinders.

Boundary Conditions A:
The appropriate conditions are

$$\left.\begin{aligned} u(\lambda R) = 0 \quad &, \quad m_{rz}(\lambda R) = 0 \\ u(R) = 0 \quad &, \quad m_{rz}(R) = 0 \end{aligned}\right\} \tag{3.6.16}$$

The general solution, given by Eqs. (3.6.3) to (3.6.6), then reduces to

$$u = -\frac{R}{4\mu}\frac{\partial p}{\partial\theta}\left[\frac{\lambda^2 \ln \lambda^2}{1 - \lambda^2}\left(\frac{R}{r} - \frac{r}{R}\right) - \frac{r}{R}\ln(r/R)^2\right] + \frac{1}{a^2}\frac{R}{\mu}\frac{\partial p}{\partial\theta}\left[\frac{R}{r} - S(r/l)\right] \tag{3.6.17}$$

$$t_{r\theta} = \frac{1}{2}\frac{\partial p}{\partial \theta}\left[1 + \left(\frac{\lambda^2 \ln \lambda^2}{1-\lambda^2}\right)\frac{R^2}{r^2}\right] - \frac{2}{a^2}\frac{R}{r}\frac{\partial p}{\partial \theta}\left[\frac{R}{r} - S(r/l)\right] \tag{3.6.18}$$

where

$$S(r/l) = \frac{Z_1 I_1(r/l) - Y_1 K_1(r/l)}{Z_1 I_1(a) - Y_1 K_1(a)}$$

$$Y_1 = I_1(a) - \lambda I_1(\lambda a) \quad , \quad Z_1 = K_1(a) - \lambda K_1(\lambda a)$$

The shear stress is affected by both μ and η. However, since $S(R/l) = S(a) = 1$ and $S(\lambda R/l) = S(\lambda a) = 1/\lambda$, the contribution of the couple stresses to the shear stress $t_{r\theta}$ drops out at the walls. Thus the wall shear is unaffected in form.

Boundary Conditions B:

The boundary conditions for this case are $u(\lambda R) = 0$, $u(R) = 0$, $\omega_z(\lambda R) = 0$ and $\omega_z(R) = 0$. The solution is then given by Eqs. (3.6.3) to (3.6.6), where the constants A_1, A_2, A_3 and A_4 are given by

$$\left.\begin{aligned}
A_1 &= \frac{R^2}{4\mu}\frac{\partial p}{\partial \theta}\left[1 - \frac{2}{aD}\left\{Z_4\left[I_1(a) - \frac{a}{2}I_0(a)\right]\right.\right.\\
&\qquad \left.\left.- Y_4\left[K_1(a) - \frac{a}{2}K_0(a)\right]\right\}\right]\\
A_2 &= -\frac{1}{4\mu}\frac{\partial p}{\partial \theta}\left[\ln R^2 + \frac{1}{D}\left(Z_4 I_0(a) + Y_4 K_0(a)\right)\right]\\
A_3 &= \frac{R}{2a\mu}\frac{\partial p}{\partial \theta}\frac{Z_4}{D} \quad , \quad A_4 = -\frac{R}{2a\mu}\frac{\partial p}{\partial \theta}\frac{Y_4}{D}
\end{aligned}\right\} \tag{3.6.19}$$

where D, Y_0, Y_1, Y_2, Z_0, Z_1 and Z_2 are given by Eq. (3.6.14), and

$$Y_4 = Y_2 \ln \lambda^2 - \frac{a}{2}(1 - \lambda^2 + \ln \lambda^2)\, Y_0$$

$$Z_4 = Z_2 \ln \lambda^2 + \frac{1}{2}(1 - \lambda^2 + \ln \lambda^2)\, Z_0$$

Thus, this flow is affected by couple stresses for both boundary conditions A and B.

3.7 Poiseuille Flow Through Circular Pipes

Consider the steady flow of an incompressible fluid through a circular pipe of radius R, due to an axial pressure gradient. In cylindrical polar coordinates (r, θ, x), the velocity field is of the form

$$v_r = 0 \quad , \quad v_\theta = 0 \quad , \quad v_x = u(r) \tag{3.7.1}$$

where x is measured along the centerline of the pipe. The equations of motion for this case can then be shown to reduce to

$$\nabla^4 u - \frac{1}{l^2} \nabla^2 u = - \frac{1}{\eta} \frac{dp}{dx} \quad , \quad p = p(x) \tag{3.7.2}$$

where $\nabla^2 u \equiv \dfrac{1}{r} \dfrac{\partial}{\partial r} \left[r \dfrac{\partial u}{\partial r} \right]$.

The general solution of this equation is given by

$$u = \frac{1}{4\mu} \frac{dp}{dx} r^2 + A_1 + A_2 \ln r + C_0(r/l) \tag{3.7.3}$$

$$\omega_\theta = - \frac{1}{2} \frac{\partial u}{\partial r} = \frac{1}{4\mu} \frac{dp}{dx} r - \frac{A_2}{2r} - \frac{1}{2l} C_0'(r/l) \tag{3.7.4}$$

$$k_{r\theta} = \frac{\partial \omega_\theta}{\partial r} = - \frac{1}{4\mu} \frac{dp}{dx} + \frac{A_2}{2r^2} - \frac{1}{2l^2} C_0''(r/l) \tag{3.7.5}$$

$$k_{\theta r} = - \frac{\omega_\theta}{r} = \frac{1}{4\mu} \frac{dp}{dx} + \frac{A_2}{2r^2} + \frac{1}{2rl} C_0'(r/l) \tag{3.7.6}$$

$$\begin{aligned} m_{r\theta} = &- (l^2 - l_1^2) \frac{dp}{dx} + 2(\eta + \eta') \frac{A_2}{r^2} \\ &- 2\mu [C_0(r/l) - (1 + l_1^2/l^2) \frac{l}{r} C_0'(r/l)] \end{aligned} \tag{3.7.7}$$

where $l_1^2 = \eta'/\mu$ and $C_0(r/l) = A_3 I_0(r/l) + A_4 K_0(r/l)$.

Also, by using $t_{rx}^S = \mu \partial u/\partial r$ and $t_{rx}^A = \frac{1}{2} [\partial m_{r\theta}/\partial r + (m_{r\theta} + m_{\theta r})/r]$,

$$t_{rx} = t_{rx}^S + t_{rx}^A = \frac{1}{2} \frac{dp}{dx} r + \frac{\mu A_2}{r} \tag{3.7.8}$$

Of the flows discussed so far, this is the first one in which the material constant η' is present in the expression for the component of the couple stress which acts at a wall.

Boundary Conditions A:

The appropriate boundary conditions are that u and ω_θ be finite at $r = 0$, and that $u(R) = 0$ and $m_{r\theta}(R) = 0$, where R is the radius of the pipe. The conditions at $r = 0$ require that $A_2 = 0$ and $A_4 = 0$. The resulting solutions are given by

$$U(\xi) = u(\xi)/\left[-\frac{R^2}{4\mu}\frac{dp}{dx}\right] = 1 - \xi^2 + \frac{2}{a^2}(1 - l_1^2/l^2)\frac{I_0(a\xi) - I_0(a)}{N(a)} \tag{3.7.9}$$

$$\Omega(\xi) = \omega_\theta(\xi)/\left[-\frac{R}{4\mu}\frac{dp}{dx}\right] = \xi - \frac{1}{a}(1 - l_1^2/l^2)\frac{I_0'(a\xi)}{N(a)} \tag{3.7.10}$$

$$m_{r\theta} = -l^2(1 - l_1^2/l^2)\frac{dp}{dx}\left[1 - \frac{N(a\xi)}{N(a)}\right] \tag{3.7.11}$$

where $\xi = r/R$, $a = R/l$, and $N(a) = I_0(a) - \frac{1}{a}(1 + l_1^2/l^2)I_0'(a)$.

The flow through $Q = \int_0^R 2\pi ru\, dr$ is given by

$$\frac{Q}{Q_0} = 1 - \frac{4}{a^3}(1 - l_1^2/l^2)\frac{aI_0(a) - 2I_0'(a)}{N(a)} \tag{3.7.12}$$

where Q_0 has the same connotation as in Section (3.5) and is given by

$$Q_0 = -\left[\frac{\pi R^4}{8\mu}\right]\frac{dp}{dx} \tag{3.7.13}$$

From Eq. (3.7.8) $t_{rx} = \frac{1}{2}(dp/dx)r$, since $A_2 = 0$, so that the wall shear is given by

$$\tau_w = \frac{1}{2}\frac{dp}{dx}R \tag{3.7.14}$$

Thus the expression for the wall shear is unaffected by couple stresses.

An interesting feature of this flow is that all the three material constants μ, η and η' are present. Since μ and η can be determined by methods described in the previous sections, this flow gives an experimental method for determining the constant η'. Another interesting feature is that all the effects of couple stresses will be absent for a material for which $\eta = \eta'$. Now $\eta = \eta'$ is equivalent to requiring that the couple stress tensor be symmetric. Thus, if the couple stress tensor is symmetric, then all its effects will be absent for pipe flow for boundary conditions A.

The variation of Q/Q_0 versus a, with $\nu = \eta'/\eta = l_1^2/l^2$ as parameter, is shown in Fig. 3.7.1. From Eq. (3.7.12), $Q/Q_0 \equiv 1$ for $\nu = 1$. $\nu = 0$

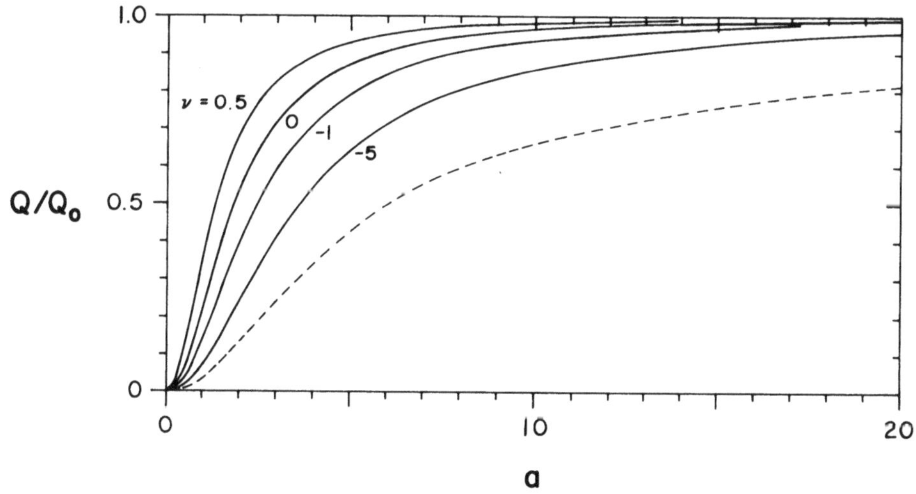

Fig. 3.7.1 Variation of Q/Q_0 versus a for pipe Poiseuille flow with boundary conditions A and B.

corresponds to the constitutive equation for couple stresses reducing to $m_{ij} = 4\eta\, k_{ij}$, the second term on the right hand side being zero. As ν decreases from one, Q/Q_0 decreases from a maximum of $Q/Q_0 = 1$.

Boundary Conditions B:

Here again u, ω_θ and $m_{r\theta}$ must be finite at $r = 0$, which requires that $A_2 = 0$ and $A_r = 0$. Furthermore, the solution must satisfy the boundary conditions $u(R) = 0$ and $\omega_\theta(R) = 0$. The general solution, given by Eqs. (3.7.3) to (3.7.8), then reduces to

$$U(\xi) = u(\xi)/\left[-\frac{R^2}{4\mu}\frac{dp}{dx}\right] = 1 - \xi^2 + \frac{2}{aI_0'(a)}[I_0(a\xi) - I_0(a)] \quad (3.7.15)$$

$$\Omega(\xi) = \omega_\theta/\left[-\frac{R}{4\mu}\frac{dp}{dx}\right] = \xi - \frac{I_0'(a\xi)}{I_0'(a)} \quad (3.7.16)$$

$$m_{r\theta} = -l^2\frac{dp}{dx}\left[(1 - l_1^2/l^2) - \frac{a}{I_0'(a)}\left\{I_0(a\xi) - (1 + l_1^2/l^2)\frac{I_0'(a\xi)}{a\xi}\right\}\right] \quad (3.7.17)$$

and $t_{rx} = \frac{1}{2}(dp/dx)r$, so that

$$\tau_w = \frac{1}{2}\frac{dp}{dx}R \quad (3.7.18)$$

The through flow Q is given by

$$\frac{Q}{Q_0} = 1 + \frac{8}{a^2} - \frac{4}{a}\frac{I_0(a)}{I_1(a)} \tag{3.7.19}$$

Notice that, for these boundary conditions, the velocity and vorticity fields are not affected by η' while $m_{r\theta}$ is.

The distribution of Q/Q_0 as a function of a is shown by a dotted line in Fig. 3.7.1. The flow Q/Q_0 is smaller for boundary conditions B than that for boundary conditions A.

3.8 Creeping Flow Past a Sphere

Creeping flow past a sphere is best studied by introducing spherical polar coordinates (r, θ, ϕ) centered on the sphere. The surface of the sphere of radius R is then given by $r = R$. The flow is assumed to be uniform at infinity, having a velocity v_∞ in the z, that is, the $\theta = 0$ direction. Because of symmetry $v_\phi = 0$. The condition at infinity can be expressed by: $v_r = v_\infty \cos\theta$ and $v_\theta = -v_\infty \sin\theta$ at infinity.

The problem is then to solve the creeping flow equation

$$\nabla p = \mu\nabla^2\mathbf{v} - \eta\nabla^4\mathbf{v} \tag{3.8.1}$$

which is obtained from the equations of motion by neglecting the inertia terms.

Boundary Conditions A:

In addition to the conditions at infinity, the desired solution of Eq. (3.8.1) must also satisfy the conditions

$$v_r(R) = 0 \quad , \quad v_\theta(R) = 0 \quad , \quad m_{r\phi}(R) = 0 \tag{3.8.2}$$

where R is the radius of the sphere. The velocity distribution is then given by

$$V_r(\xi) = \frac{v_r(\xi)}{v_\infty} = \left\{\left(1 - \frac{3}{2\xi} + \frac{1}{2\xi^3}\right) - \frac{3\alpha}{2}\left[\frac{1}{a\xi} - \left(\frac{1}{a} + \frac{2}{a^2} + \frac{2}{a^3}\right)\frac{1}{\xi^3}\right.\right.$$
$$\left.\left. + \frac{2}{a^2\xi^2}\left(1 + \frac{1}{a\xi}\right)\exp(-a(\xi - 1))\right]\right\}\cos\theta \tag{3.8.3}$$

$$V_\theta(\xi) = \frac{v_\theta(\xi)}{v_\infty} = -\left\{\left(1 - \frac{3}{4\xi} - \frac{1}{4\xi^3}\right) - \frac{3\alpha}{4}\left[\frac{1}{a\xi} + \left(\frac{1}{a} + \frac{2}{a^2} + \frac{2}{a^3}\right)\frac{1}{\xi^3}\right.\right.$$

$$- \frac{2}{a\xi}\left(1 + \frac{1}{a\xi} + \frac{1}{a^2\xi^2}\right)\exp(-a(\xi - 1))\Big]\Big\}\sin\theta \tag{3.8.4}$$

where $a = R/l$, $\xi = r/R$, $a\xi = r/l$ and $\alpha = (2 + \nu)/(a + 2 + \nu)$.

The pressure drag, D_n, and the shear drag, D_s, are given by

$$D_n = 2\pi\mu R v_\infty (1 + \alpha/a) \tag{3.8.5}$$

$$D_s = 4\pi\mu R v_\infty (1 + \alpha/a) \tag{3.8.6}$$

where $\alpha = (2 + \nu)/(a + 2 + \nu)$ and $\nu = \eta'/\eta = l_1^2/l^2$. The total drag is then given by

$$D = 6\pi\mu R v_\infty (1 + \alpha/a) \tag{3.8.7}$$

In the nonpolar case the drag, D_0, is given by $D_0 = 6\pi\,\mu R\, v_\infty$, so that finally

$$\frac{D}{D_0} = \frac{D}{6\pi\mu R v_\infty} = 1 + \frac{\alpha}{a} \tag{3.8.8}$$

From this expression the effect of couple stresses can be seen to be equivalent to an apparent increase in the viscosity by a factor $(1 + \alpha/a)$. Notice that the drag, besides depending on η, also depends on η'.

Boundary Conditions B:

The boundary conditions on the sphere for this case are: $v_r(R) = 0$, $v_\theta(R) = 0$ and $\omega_\phi(R) = 0$. The solution then comes out to be

$$V_r(\xi) = \frac{v_r(\xi)}{v_\infty} = \Big[1 - \frac{3}{2}\left(\frac{1 + a}{a}\right)\frac{1}{\xi} + \left(\frac{1 + 3\beta}{2}\right)\frac{1}{\xi^3}$$

$$- 3\left(\frac{1}{a^2\xi^2} + \frac{1}{a^3\xi^3}\right)\exp(-a(\xi - 1))\Big]\cos\theta \tag{3.8.9}$$

$$V_\theta(\xi) = \frac{v_\theta(\xi)}{v_\infty} = -\Big[1 - \frac{3}{4}\left(\frac{1 + a}{a}\right)\frac{1}{\xi} - \left(\frac{1 + 3\beta}{4}\right)\frac{1}{\xi^3}$$

$$+ \frac{3}{2}\left(\frac{1}{a\xi} + \frac{1}{a^2\xi^2} + \frac{1}{a^3\xi^3}\right)\exp(-a(\xi - 1))\Big]\sin\theta \tag{3.8.10}$$

$$\omega_\phi(\xi) = -\frac{3v_\infty}{4R}\left[\left(\frac{1+a}{a}\right)\frac{1}{\xi^3} - a\left(\frac{1}{a\xi} + \frac{1}{a^2\xi^2}\right)\exp(-a(\xi-1))\right]\sin\theta \tag{3.8.11}$$

where $\beta = \frac{1}{a} + \frac{2}{a^2} + \frac{2}{a^3}$. The drag is given by

$$\frac{D}{D_0} = 1 + \frac{1}{a} \tag{3.8.12}$$

In this case the drag does not depend on η'.

The variation of D/D_0 versus a is shown in Fig. 3.8.1 by a solid line. The dotted lines show the drag for boundary conditions A, for different values of $\nu = \eta'/\eta$, as per Eq. (3.8.8).

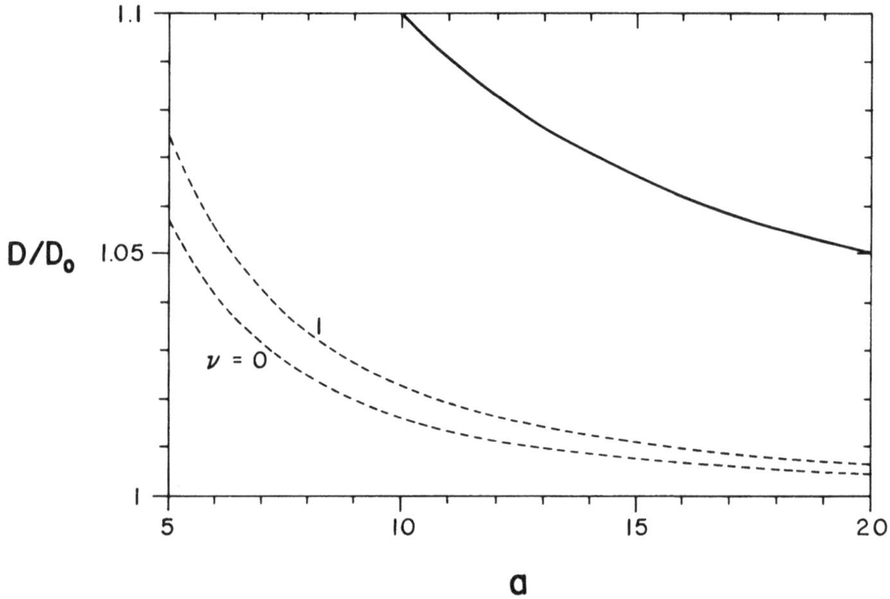

Fig. 3.8.1 Variation of D/D_0 versus a for creeping flow past a sphere with boundary conditions A and B.

3.9 Some Time-Dependent Flows

Here only parallel flow fields between two plates will be considered. The velocity field may then be assumed to be $v_x = u(y,t)$, $v_y = 0$ and $v_z = 0$, where the coordinate y is normal to the plates, x is along the flow direction, and t is the time. For such fields, the equations of motion reduce to

$$\frac{\partial u}{\partial t} = -\frac{1}{\rho}\frac{\partial p}{\partial x} + \frac{\mu}{\rho}\frac{\partial^2 u}{\partial y^2} - \frac{\eta}{\rho}\frac{\partial^4 u}{\partial y^4}, \quad p = p(x,t) \tag{3.9.1}$$

Introducing the nondimensional variables $\xi = y/h$ and $\tau = \mu t/\rho h^2$, where h is the distance between the parallel plates, Eq. (3.9.1) can be rewritten as

$$\frac{\partial u}{\partial \tau} = -\frac{h^2}{\mu}\frac{\partial p}{\partial x} + \frac{\partial^2 u}{\partial \xi^2} - \frac{1}{a^2}\frac{\partial^4 u}{\partial \xi^4} \tag{3.9.2}$$

where $a = h/l$.

The solutions for the time-dependent case are only available for boundary conditions A. Solutions will only be presented for a few simple cases to illustrate the effect of couple stresses on the time constants of flow.

1. Development of Plane Couette Flow

In this case the flow is started from rest at time $t = 0$ by the sudden acceleration of the upper plate which subsequently moves with a constant velocity V. The pressure gradient is assumed to be zero throughout. The development of flow is then governed by Eq. (3.9.2) with $\partial p/\partial x = 0$. For boundary conditions A, the appropriate initial and boundary conditions are: $u(\tau,\xi) = 0$ for $\tau \leqslant 0$ and $0 \leqslant \xi \leqslant 1$; and $u(0,\tau) = 0$, $u(1,\tau) = V$, $m_{\xi z}(0,\tau) = 0$, and $m_{\xi z}(1,\tau) = 0$ for $\tau > 0$. The solution satisfying these conditions is

$$U(\xi,\tau) = \frac{u(\xi,\tau)}{V} = \xi - \sum_{m=1}^{\infty} C_m \exp(-\lambda^2\tau)\sin m\pi\xi \tag{3.9.3}$$

where $\lambda^2 = m^2\pi^2(1 + m^2\pi^2/a^2)$ and $C_m = 2(-1)^{m+1}/m\pi$. The shear stress $t_{yx} = t_{yx}^S + t_{yx}^A$ is given by

$$\frac{ht_{yx}}{\mu V} = 1 - \sum_{m=1}^{\infty} m\pi(1 + m^2\pi^2/a^2)C_m\exp(-\lambda^2\tau)\cos m\pi\xi \tag{3.9.4}$$

The temporal development of the flow pattern, shown in Figs. 3.9.1, 3.9.2 and 3.9.3 for $a = 0.5$, 2, 5, 10, 25 and 50, where time is measured by the parameter $T = 4(\mu t/\rho)^{1/2}/h$, shows that couple stresses accelerate the development of flow. In order to obtain an estimate of this effect, a time constant T_c is introduced and is defined as the value of the time parameter $T = 4(\mu t/\rho)^{1/2}/h$ at which the velocity at the center, $\xi = 0.5$, attains 0.99 of the steady state value. T_c^0 is the value of T_c corresponding to the nonpolar case. The variations of T_c and T_c/T_c^0, with a, shown in Fig. 3.9.4, indi-

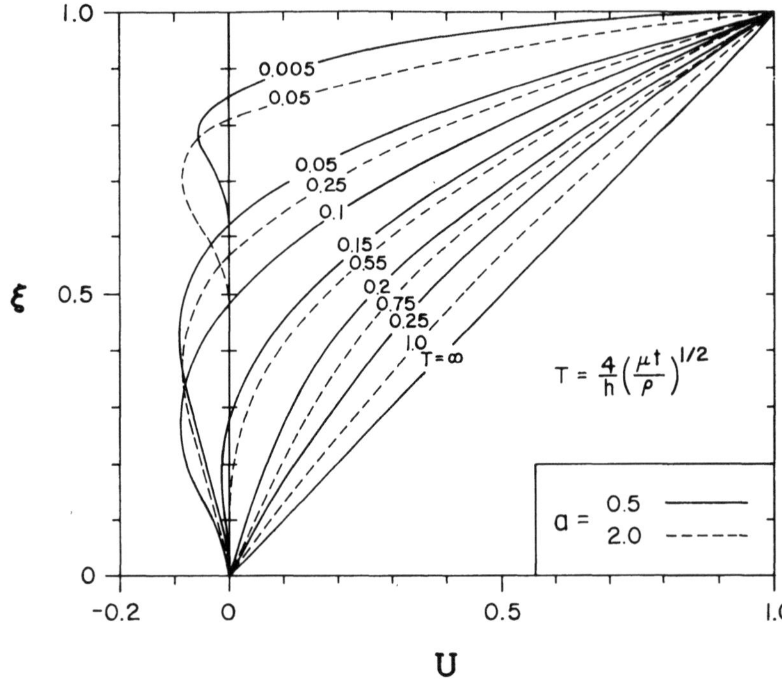

Fig. 3.9.1 Development of plane Couette flow, for $a = 0.5$ and $a = 2$, with boundary conditions A. *Adapted from Ref. [6].*

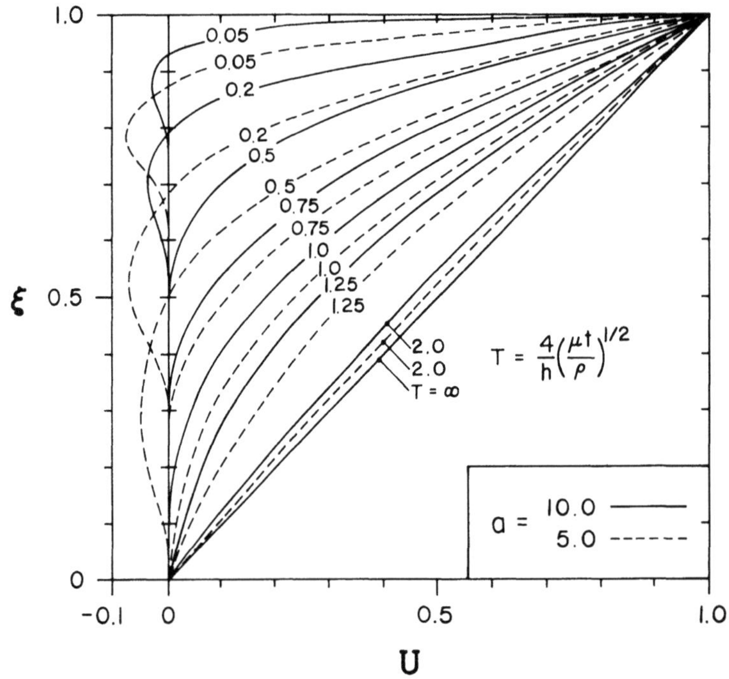

Fig. 3.9.2 Development of plane Couette flow, for $a = 5$ and $a = 10$, with boundary conditions A. *Adapted from Ref. [6].*

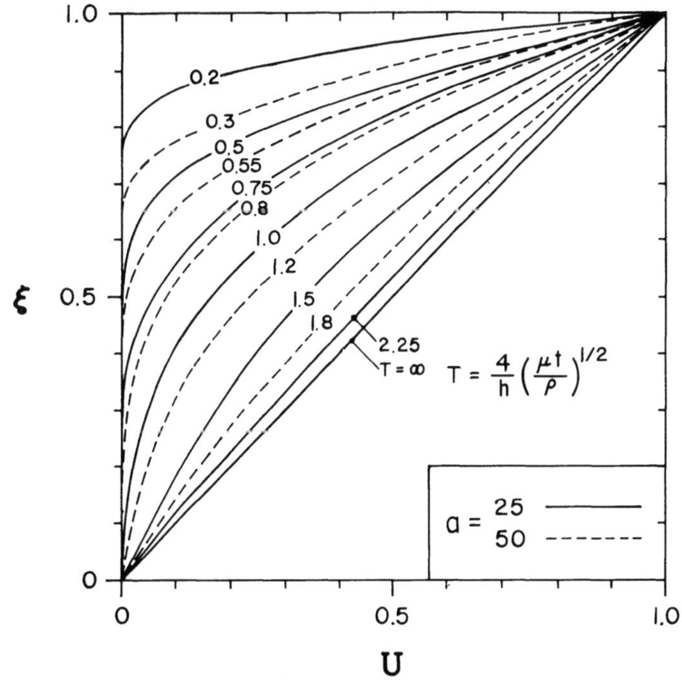

Fig. 3.9.3 Development of plane Couette flow, for $a = 25$ and $a = 50$, with boundary conditions A. *Adapted from Ref. [6].*

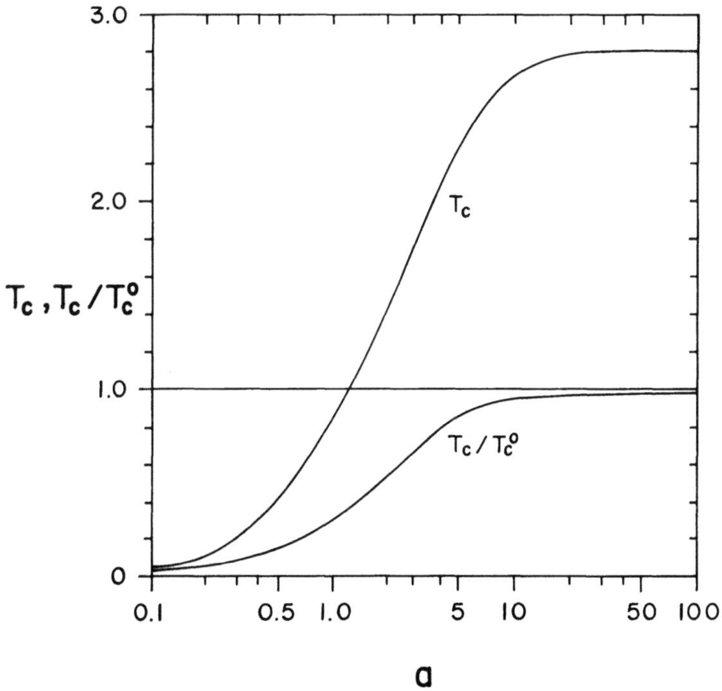

Fig. 3.9.4 Variation of the time constant with a. *Adapted from Ref. [6].*

cate that for low values of a the time constant is reduced appreciably and that for couple stresses approaching infinity, or $a \to 0$, the steady state linear profile develops instantaneously.

For low values of the parameter a, a peculiar behavior of reverse flow in the initial stages of the development of flow is predicted, as can be seen from Figs. 3.9.1 and 3.9.2. There is the possibility that this behavior may not occur if boundary conditions B are used. This effect is not present for larger values of a. The limiting case of zero couple stresses, $a \to \infty$, is the same as the classical nonpolar solution. In general, the wall shear is larger for larger values of couple stresses and approaches the steady state value faster.

2. Development of Plane Poiseuille Flow

Consider the development of the flow between two stationary parallel plates, distance h apart, due to the imposition of a constant axial pressure gradient $\partial p/\partial x$. For boundary conditions A, the appropriate initial and boundary conditions are: $u(\xi,\tau) = 0$ for $\tau \leqslant 0$ and $0 \leqslant \xi \leqslant 1$; and $u(0,\tau) = 0$, $u(1,\tau) = 0$, $m_{\xi z}(0,\tau) = 0$ and $m_{\xi z}(1,\tau) = 0$ for $\tau > 0$. The velocity is given then by

$$U(\xi,\tau) = \frac{u(\xi,\tau)}{-\dfrac{h^2}{2\mu}\dfrac{\partial p}{\partial x}} = \frac{1}{2}\xi(1-\xi) + \frac{1}{a^2}\left[\frac{\sinh a\xi + \sinh a(1-\xi)}{\sinh a} - 1\right] + \sum_{m=1,3,5,\ldots}^{\infty} D_m \exp(-\lambda^2\tau)\sin m\pi\xi \tag{3.9.5}$$

where $\lambda^2 = m^2\pi^2(1 + m^2\pi^2/a^2)$, as before, and $D_m = 4/[(m\pi)^3(1+m^2\pi^2/a^2)]$, $m = 1, 3, 5, \cdots$, and the shear stress is given by

$$\frac{t_{yx}}{-h\dfrac{\partial p}{\partial x}} = \frac{1}{2}(1-\xi) + \sum_{m=1,3,5,\ldots}^{\infty} (4/m^2\pi^2)\exp(-\lambda^2\tau)\cos m\pi\xi \tag{3.9.6}$$

The development of the flow profiles is shown in Figs. 3.9.5 and 3.9.6 for a equal to 1, 2, 5 and 10. For values of a higher than about 20, the effects of couple stresses are negligible. Here again, the flow accelerates to the steady state value faster than in the nonpolar case. Thus the effect of couple stresses is roughly equivalent to an apparent increase in the viscosity of the fluid. If the same time constant T_c, as defined for Couette flow, is used again, then the numerical values of T_c are about the same as for

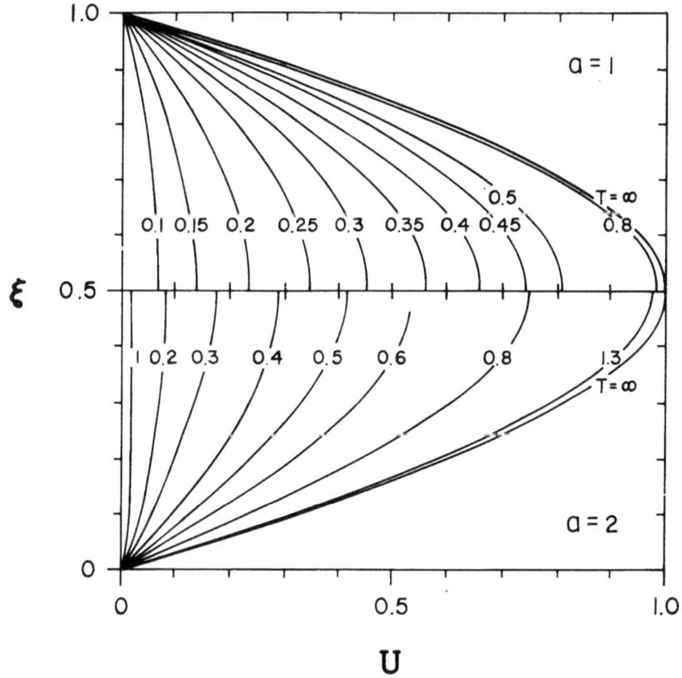

Fig. 3.9.5 Development of plane Poiseuille flow, for $a = 1$ and $a = 2$, with boundary conditions A. *Adapted from Ref. [6].*

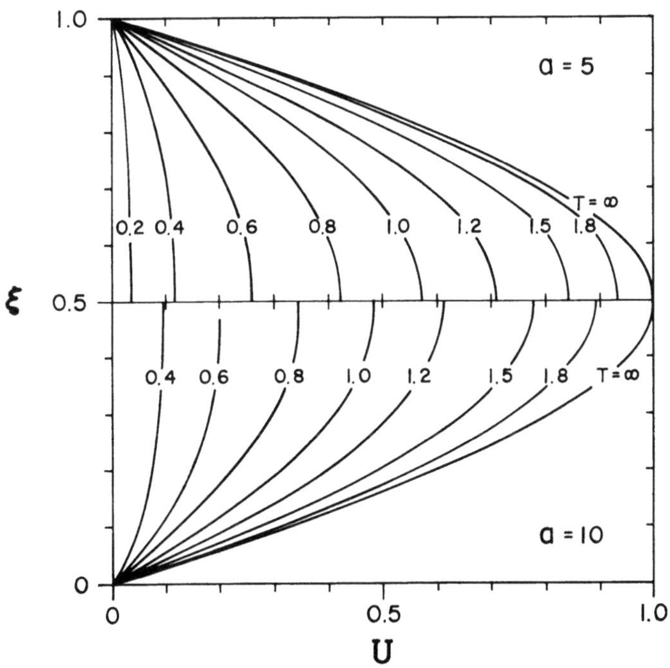

Fig. 3.9.6 Development of plane Poiseuille flow, for $a = 5$ and $a = 10$, with boundary conditions A. *Adapted from Ref. [6].*

Couette flow. The wall shear is again higher for larger values of couple stresses.

The solutions presented in this section are due to Usmani [Ref. 6], where solutions for the following time-dependent flows are also given: flow due to a harmonically oscillating plate, pulsatile flow between parallel plates, and the development of cylindrical Couette flow.

3.10 Stability of Plane Poiseuille Flow

Some of the effects of couple stresses on the stability of plane Poiseuille flow, between two parallel plates, are described in this section. Only boundary conditions A are considered. Following the usual assumptions of linearized stability theory, the nondimensional velocity $v_i(x_i,t)$ is assumed to consist of a steady state value $\overline{v_i}(x_i)$, whose stability constitutes the subject of investigation, and a small disturbance v_i' such that $v_i(x_i,t) = \overline{v_i}(x_i) + v_i'(x_i,t)$. Similarly, the nondimensional pressure is resolved as $p(x_i,t) = \overline{p}(x_i) + p'(x_i,t)$. Then the linearized equation, which governs the variation of the disturbances, can be shown to be

$$\frac{\partial v_i'}{\partial t} + \overline{v}_r v_{i,r}' + v_r'\overline{v}_{i,r} = -p_{,i}' + \frac{1}{R}\left[v_{i,rr}' - \frac{1}{a^2} v_{i,rrss}'\right] \tag{3.10.1}$$

where $R = \rho^* v^*_{\max} h/\mu$ is the Reynolds number, the quantities with asterisks representing actual dimensional physical quantities. $v^*_{\max}$ is some representative velocity, which in this case will be the maximum velocity of the steady state profile. Barred letters represent the nondimensional values of the steady state motion. Letters without bars represent nondimensional quantities and primed quantities represent nondimensional disturbances. The nondimensional quantities are defined as follows: $x_i = x^*_i/h$, $v_i = v^*_i/v^*_{\max}$, $t = t^*v^*_{\max}/h$ and $p = p^*/\rho^* v^{*2}_{\max}$ where $2h$ is the distance between the two plates, and $v^*_{\max}$ is the maximum velocity of the steady flow.

The stream function for the disturbance, ψ', such that $u' = -\partial\psi'/\partial y$, $v' = -\partial\psi'/\partial x$, may be assumed have the form

$$\psi'(x,y,t) = \phi(y) \exp[i\alpha(x - ct)] \tag{3.10.2}$$

where α is a real quantity denoting the wave number of the disturbance in the x direction. ψ', ϕ and c are complex quantities. Only the real part of ψ' gives the actual stream function. Substitution from Eq. (3.11.2) in Eq. (3.10.1) a, b, and an elimination of p' from the resulting equations then gives the linearized disturbance equation

$$(D^2 - \alpha^2)^2\left[1 - \frac{1}{a^2}(D^2 - \alpha^2)\right]\phi = i\alpha R\left[(\bar{u} - c)(D^2 - \alpha^2)\psi - (D^2\bar{u})\psi\right] \tag{3.10.3}$$

where $D \equiv d/dy$. For boundary conditions A, the conditions on ϕ are $\phi = 0$, $D\phi - 0$ and $(D^2 - \alpha^2)d\phi = 0$ at $y = 0$ and $y = 2$.

As $a \rightarrow \infty$, that is, as couple stresses vanish, Eq. (3.10.3) reduces to the well-known Orr-Sommerfeld equation

$$(D^2 - \alpha^2)^2\phi = i\alpha R[(\bar{u} - c)(D^2 - \alpha^2)\phi - (D^2\bar{u})\phi]$$

Equation (3.10.3), subject to the boundary conditions given above, results in an eigenvalue problem of the type $F(\alpha,R,a,c) = 0$. If the imaginary part c_i of c is positive, the disturbance is unstable. If $c_i < 0$, the disturbance is damped in time. If $c_i = 0$, there is a sustained oscillation. The condition $c_i = 0$ gives rise to a neutral stability curve in the $\alpha-R$ plane for each fixed value of the couple stress parameter a.

Neutral stability curves for plane Poiseuille flow, for different values of a, are shown in Fig. 3.10.1. The dotted curve represents the solution of the Orr-Sommerfeld equation for the nonpolar case and, therefore, corresponds

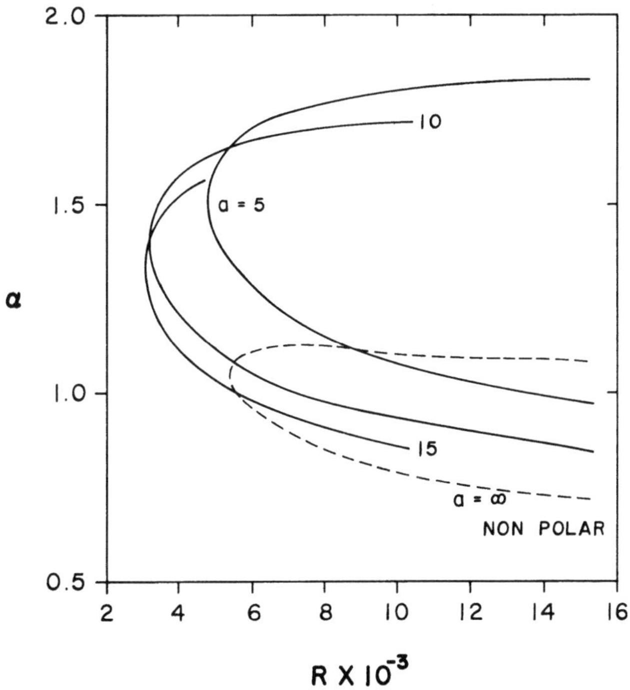

Fig. 3.10.1 Neutral stability curves for plane Poiseuille flow for a = 5, 10, 15 and ∞ with boundary conditions A. *Adapted from Ref. [8].*

to the case $a \rightarrow \infty$. The minimum Reynolds number on each curve represents the critical Reynolds number R_{cr}, below which all small disturbances are damped. Figure 3.10.2 shows the variation of R_{cr} with a. From the computations, R_{cr} is available only for $a = 5$, 10 and 15. As couple stresses increase, that is, as a increases, the value of R_{cr} first decreases and then increases after passing through a minimum in the neighborhood of a lying between 10 and 15. Since for $a = 0$ the steady state velocity is identically zero, $R_{cr} \rightarrow \infty$ as $a \rightarrow 0$. Figure 3.10.1 shows that the domain of the unstable region increases appreciably in the presence of couple stresses.

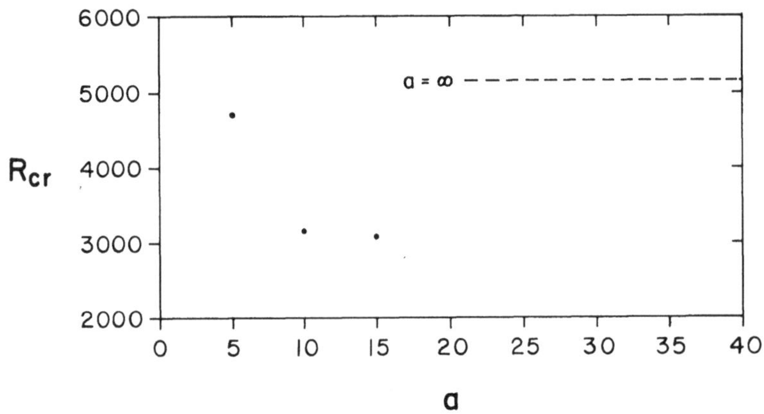

Fig. 3.10.2 Variation of the critical Reynolds number R_{cr} with a. *Adapted from Ref. [8].*

3.11 Hydromagnetic Channel Flows

Couple stresses, for the model under consideration, depend on vorticity gradients. Since vorticity gradients are known to be large in hydromagnetic flows of nonpolar fluids, couple stress effects may be expected to be large in electrically conducting polar fluids also.

Parallel steady flows of the type where the velocity is of the form $v_1 = u(z)$, $v_2 = 0$, $v_3 = 0$, will be considered. These flows are assumed to be subjected to a uniform magnetic field B_o in the positive z direction. Because of the fluid motion, a constant electric field E_o in the y direction and a magnetic field B_x in the x direction will be generated. If Gaussian units are used, then, after making the usual assumptions of magnetohydrodynamics such as, for example, neglecting the displacement current, the equations of motion reduce to

$$0 = -\frac{\partial p}{\partial x} + \frac{\sigma B_o}{c}\left[E_o - \frac{B_o}{c}\right]u + \mu\frac{d^2u}{dz^2} - \eta\frac{d^4u}{dz^4}$$

$$0 = -\frac{\partial p}{\partial y}\ , \quad 0 = -\frac{\partial p}{\partial z} - \frac{\sigma B_x}{c}\left[E_o - \frac{B_o}{c}u\right]$$

where σ is the conductivity of the fluid, and c the velocity of light.

The effects of couple stresses enter only through the last term in the first equation. Only the first equation need be considered as it determines the velocity distribution completely. Nondimensionalizing the distance z through $\zeta = z/h$, this equation can be written as

$$\frac{d^4u}{d\zeta 4} - a^2\frac{d^2u}{d\zeta^2} + a^2M^2u = a^2\left[\frac{h^2}{\mu}\ P + \frac{cE_o}{B_o}\ M^2\right] \tag{3.11.1}$$

where $a = h/l$, $P = -\partial p/\partial x$ and $M = (B_o h/c)(\sigma/\mu)^{1/2}$ is the Hartmann number.

In this section, only boundary conditions A will be considered.

A. Hydromagnetic Couette Flow

Consider the flow between two parallel plates distance h apart, when the top plate moves with a velocity V in the positive x direction. For boundary conditions A, the appropriate boundary conditions for Eq. (3.11.1) are then $u(0) = 0$, $u(1) = V$, and that the couple stresses vanish at both the plates, which in turn, implies that $u''(0) = 0$ and $u''(1) = 0$. These four boundary conditions determine $u(\zeta)$ in terms of B_o and E_o. E_o is determined by assuming that the total current in the y direction is zero, that is $\int_0^h J_y\ dz = 0$, where $J_y = \sigma(E_o - B_o u/c)$ is the component of the current density in the y direction. This condition results in $cE_o/B_o = \frac{1}{2}\ V$. Subject to these conditions the nondimensional velocity $U(\zeta) = u(\zeta)/V$ is then given by

$$U_1 = \frac{1}{2}\left\{1 + \frac{1}{\alpha^2 - \beta^2}\left[b^2\ \frac{\sinh \alpha(1-\zeta) - \sinh a\zeta}{\sinh \alpha} - \alpha^2\ \frac{\sinh \beta(1-\zeta) - \sinh \beta\zeta}{\sinh \beta}\right]\right\} \tag{3.11.2}$$

$$U_2 = \frac{1}{2}\left\{1 - (1 + D\zeta)\cosh b\zeta + \left[1 + (1 + D)\cosh b - \frac{b}{2}\ (1-\zeta)\sinh b\right]\frac{\sinh b\zeta}{\sinh b}\right\} \tag{3.11.3}$$

$$U_3 = \frac{1}{2}\Bigg\{1 + \Bigg[-\cos\delta\zeta + \Bigg(\frac{\sin\delta - r\sinh\gamma}{\cosh\gamma - \cos\delta}\Bigg)\sinh\delta\zeta\Bigg]\cosh\gamma\zeta$$

$$+ \Bigg[\Bigg(\frac{\sinh\gamma + R\sin\delta}{\cosh\gamma - \cos\delta}\Bigg)\cos\delta\zeta + R\sin\delta\zeta\Bigg]\sinh\gamma\zeta\Bigg\} \tag{3.11.4}$$

where the subscripts 1, 2 and 3 on U denote the three cases: $2M/a < 1$, $2M/a = 1$ and $2M/a > 1$, respectively, $D = \frac{b}{2}(1 + \cosh b)/\sinh b$, $R = [(4M^2/a^2) - 1]^{-1/2}$, and

$$\alpha = a\left[\frac{1 + (1 - 4M^2/a^2)^{1/2}}{2}\right]^{1/2}, \quad \beta = a\left[\frac{1 - (1 - 4M^2/a^2)^{1/2}}{2}\right]^{1/2}$$

$$\gamma = \frac{a}{2}(2M/a + 1)^{1/2}, \quad \delta = \frac{a}{2}(2M/a - 1)^{1/2} \quad \text{and} \quad b = a/\sqrt{2}$$

Now $\gamma \to 0$, $\delta \to 0$ and $R \to 0$ as couple stresses become infinite, that is, as $a \to 0$. A limiting process shows that $\underset{a \to 0}{\text{limit}}\, U = \underset{a \to 0}{\text{limit}}\, U_3 = \zeta$. But this is the velocity profile that is obtained when the couple stresses are zero ($a = \infty$) and there is no magnetic field ($M = 0$). Thus in this case, higher couple stresses tend to oppose the distortion in the velocity profile that is caused by the magnetic field. Note that, for boundary conditions A, couple stresses do not have any effect on the velocity profile in the absence of a magnetic field.

Figures 3.11.1 and 3.11.2 show how the velocity profile is distorted for different values of a for Hartmann numbers of 10 and 100, respectively. For a Hartmann number of one, the velocity profile is almost indistinguishable from $U = \zeta$ for any value of a.

2. *Hydromagnetic Plane Poiseuille Flow*

Consider the steady flow due to a pressure gradient, $P = -\partial p/\partial x$, between two parallel plates distance $2h$ apart. Let the x axis coincide with the centerline. For boundary conditions A, the appropriate boundary conditions are $u(\pm 1) = 0$ and $u''(\pm 1) = 0$. Here again the ratio E_o/B_o is determined by imposing the condition that the total current in the y direction be zero, that is $\int_{-h}^{+h} J_y dz = 0$. The velocity distribution is then given by

$$U_1 = 2\left\{\frac{1 - \left[1/(\alpha^2 - \beta^2)\right]\left[\alpha^2\cosh\beta\zeta/\cosh\beta - \beta^2\cosh\alpha\zeta/\cosh\alpha\right]}{\left[M^2/(\alpha^2 - \beta^2)\right]\left[(\alpha^2/\beta)\tanh\beta - (\beta^2/\alpha)\tanh\alpha\right]}\right\} \tag{3.11.5}$$

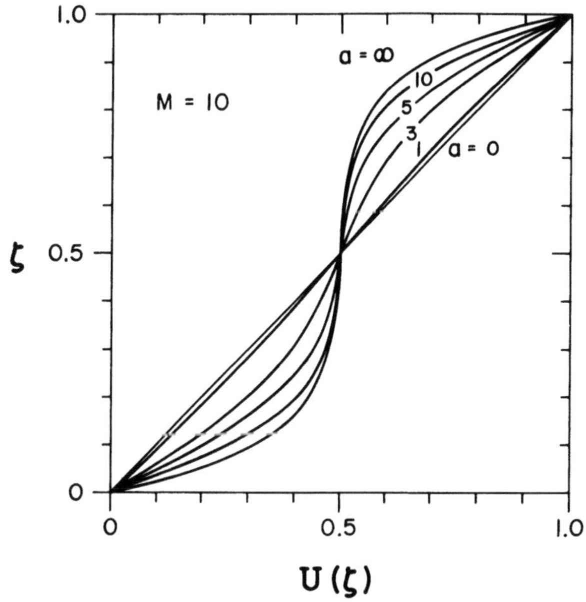

Fig. 3.11.1 Variation of the velocity profile $U(\zeta)$ for plane hydromagnetic Couette flow for Hartmann number $M = 10$. *Adapted from Ref. [4].*

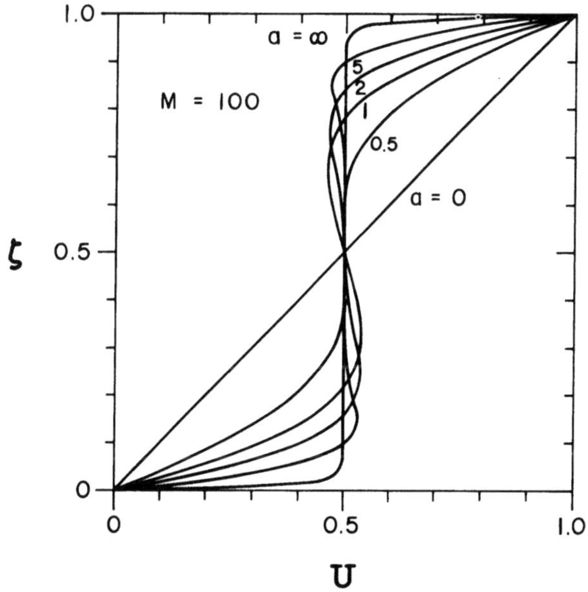

Fig. 3.11.2 Variation of the velocity profile $U(\zeta)$ for plane hydromagnetic Couette flow for Hartmann number $M = 100$. *Adapted from Ref. [4].*

$$U_2 = \frac{8}{b}\left\{\frac{1-(1+\frac{1}{2}b\tanh b)(\cosh b\zeta/\cosh b)+\frac{1}{2}b\zeta(\sinh b\zeta/\sinh b)}{(3+b\tanh b)\tanh b - b}\right\} \tag{3.11.6}$$

$$U_3 = \frac{4a}{M}\left\{\frac{(\cosh^2\gamma - \sin^2\delta) - (C\cos\delta\zeta\cosh\gamma\zeta + D\sin\delta\zeta\sinh\gamma\zeta)}{(\gamma+\delta R)\sinh 2\gamma + (\delta-\gamma R)\sin 2\delta}\right\} \tag{3.11.7}$$

where $C = \cos\delta\cosh\gamma + R\sin\delta\sinh\gamma$, $D = \sin\delta\sinh\gamma - R\cos\delta\cosh\gamma$, and $U = u/(h^2P/2\mu)$.

The velocity profiles for Hartmann numbers of 10 and 100 are shown in Figs. 3.11.3 and 3.11.4, respectively, for different values of the parameter a. Figure 3.11.4 shows that for large Hartmann numbers (here 100) the effect of couple stresses is relatively large. For $a = 100$, for no magnetic field, the maximum velocity at $\zeta = 0$ is diminished by about 0.02%, while, for $a = 100$ and $M = 100$, the maximum velocity is reduced by about 13%. This effect is quite significant. Another effect of couple stresses is the "bump" in the velocity profile near $\zeta = 1$.

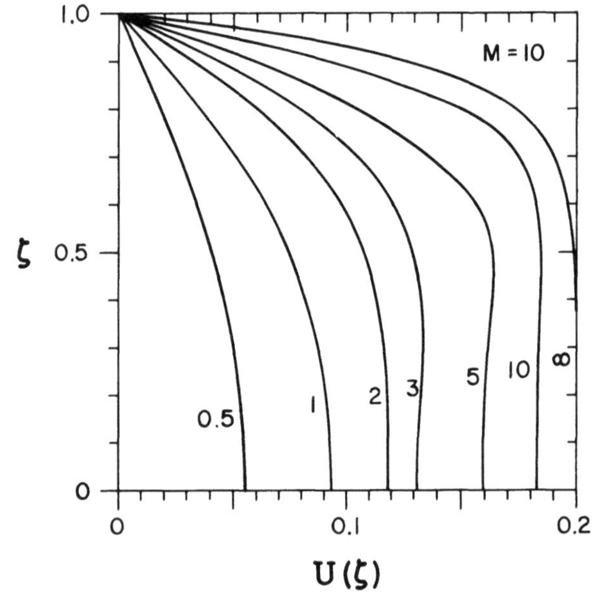

Fig. 3.11.3 Variation of the velocity profile $U(\zeta)$ for plane hydromagnetic Poiseuille flow for Hartmann number $M = 10$. *Adapted from Ref. [4].*

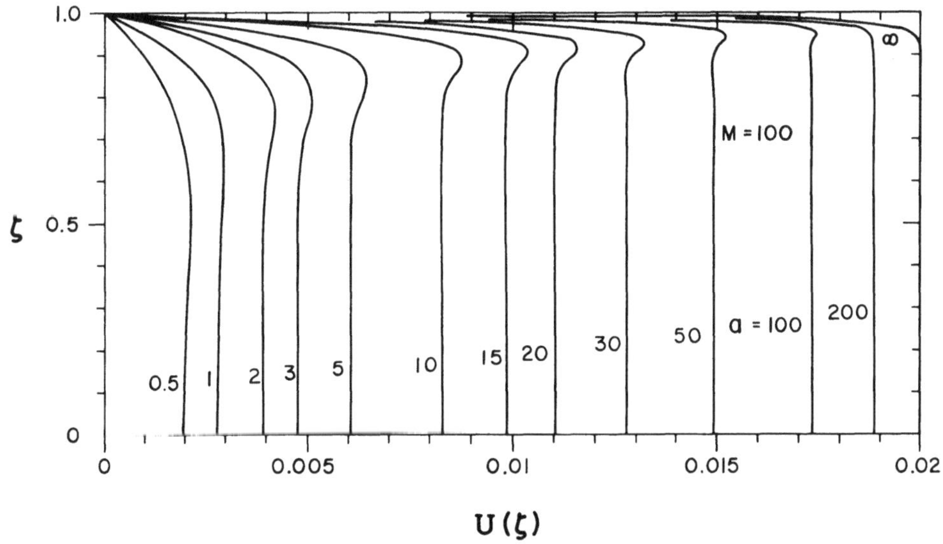

Fig. 3.11.4 Variation of the velocity profile $U(\zeta)$ for plane hydromagnetic Poiseuille flow for Hartmann number $M = 100$. *Adapted from Ref. [4].*

The flow through per unit depth, $Q = 2h \int_0^1 u(\zeta)\, d\zeta$, is related to the flow through in the absence of couple stresses and magnetic field, $Q_0 = \frac{2}{3}(h^3/\mu)P$, by the equations

$$\frac{Q_1}{Q_0} = \frac{3}{M^2}\left[\frac{1-F}{F}\right], \quad \frac{Q_2}{Q_0} = \frac{6}{b^2}\left[\frac{1-G}{G}\right], \quad \frac{Q_3}{Q_0} = \frac{3}{M^2}\left[\frac{1-H}{H}\right] \tag{3.11.8}$$

where

$$F = \frac{1}{(\alpha^2 - \beta^2)}\left[\frac{\alpha^2}{\beta}\tanh\beta - \frac{\beta^2}{\alpha}\tanh\alpha\right]$$

$$G = \frac{1}{2b}\left[(3 + b\tanh b)\tanh b - b\right]$$

$$H = \frac{(\gamma + \delta R)\sinh 2\gamma + (\delta - \gamma R)\sinh 2\delta}{2(\gamma^2 + \delta^2)(\cosh^2\gamma - \sin^2\delta)}$$

The variation of Q/Q_0 with a is shown in Fig. 3.11.5, for Hartmann numbers of 1, 10 and 100 (note the change of scale). Here again the effect of couple stresses is quite significant. For $M = 100$, the reduction in flow is quite large even for $a = 200$.

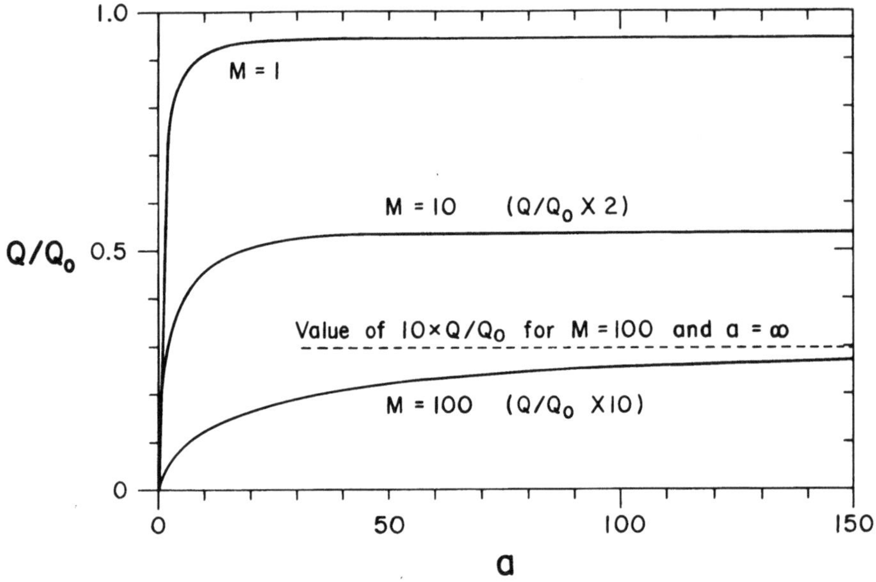

Fig. 3.11.5 The graph of Q/Q_0 versus a for Hartmann numbers of 1, 10 and 100. *Adapted from Ref. [4].*

Thus magnetic fields would seem to "magnify" the effects of couple stresses in that "observable" effects are predicted for much larger values of a. For example, for $a = 100$ the effects of couple stresses are negligible in the absence of a magnetic field, whereas for $a = 100$ and $M = 100$, the reduction in the maximum velocity is about 13% and the reduction in the flow about 14%.

3.12 Some Effects on Heat Transfer

Some aspects of the structure of the energy equation can be studied by considering the heat transfer in flow between two parallel plates.

Substitution from the constitutive equations, given in Eqs. (3.2.3) and (3.2.4), in the energy equation, given in Eq. (2.8.8), gives

$$\rho\dot{\epsilon} = -p(v_{r,r}) + \lambda(v_{r,r})^2 + \mu(v_{r,s}v_{r,s} + v_{r,s}v_{s,r}) + 4\eta\omega_{r,s}\omega_{r,s} + 4\eta'\omega_{r,s}\omega_{s,r} - q_{r,r} + \rho h \tag{3.12.1}$$

where the spin kinetic energy density k_s has been assumed to be zero.

Considering only thermal energy, $\dot{\epsilon} = c\dot{T}$ where c is the specific heat of the fluid, assumed to be a constant, and T is the absolute temperature.

Moreover $q_r = -kT_{,r}$ when only thermal flux of energy is considered. The energy equation may then be written as

$$\rho c \dot{T} = -p(v_{r,r}) + \lambda(v_{r,r})^2 + \mu(v_{r,s} v_{r,s} + v_{r,s} v_{s,r})$$
$$+ 4\eta \omega_{r,s} \omega_{r,s} + 4\eta' \omega_{r,s} \omega_{s,r} + kT_{,rr} + \rho h \qquad (3.12.2)$$

Finally, in the absence of a heat source for incompressible fluids, the above equation reduces to

$$\rho c \dot{T} = \mu(v_{r,s} v_{r,s} + v_{r,s} v_{s,r})$$
$$+ 4\eta \omega_{r,s} \omega_{r,s} + 4\eta' \omega_{r,s} \omega_{s,r} + kT_{,rr} \qquad (3.12.3)$$

Flow Between Parallel Plates

For flows of the type $v_1 = u(y)$, $v_2 = 0$ and $v_3 = 0$, Eq. (3.12.3) reduces to

$$\rho c \left[\frac{\partial T}{\partial t} + u \frac{\partial T}{\partial x} \right] = \mu \left[\frac{\partial u}{\partial y} \right]^2 + 4\eta \left[\frac{\partial \omega_z}{\partial y} \right]^2 + k \frac{\partial^2 T}{\partial y^2} \qquad (3.12.4)$$

so that for steady flow

$$-\frac{\partial^2 T}{\partial y^2} = \frac{\mu}{k} \left[\frac{\partial u}{\partial y} \right]^2 + \frac{4\eta}{k} \left[\frac{\partial \omega_z}{\partial y} \right]^2 \qquad (3.12.5)$$

The effects of couple stresses can be seen to enter directly through the second term on the right-hand side, and indirectly through the dependence of the velocity on the couple stresses.

Only boundary conditions A will be considered in this section.

1. Couette Flow

The discussion in Section (3.5) has shown that couple stresses are absent for boundary conditions A, and, since $\partial \omega_z / \partial y = 0$ for this flow, the temperature distribution will be the same as in the nonpolar case. However, the temperature distribution will be affected for boundary conditions B.

2. Plane Poiseuille Flow

The results for this case will also be presented for boundary conditions A only. Consider the steady plane Poiseuille flow discussed in Section (3.5). Let the plates at $y = \pm h$, that is, at $\xi = \pm 1$, be maintained at a constant temperature T_0. Also let $\theta = T - T_0$. Then the energy equation, given in Eq. (3.12.5), reduces to

$$-\frac{d^2\theta}{d\xi^2}=\frac{\mu}{k}\left[\left(\frac{du(\xi)}{d\xi}\right)^2+4l^2\left(\frac{d\omega_z(\xi)}{d\xi}\right)^2\right] \tag{3.12.6}$$

which, upon substitution for $u(\xi)$ and $\omega_z(\xi)$ from Eqs. (3.5.17) and (3.5.18), becomes

$$\frac{d^2\theta}{d\xi^2}=-\frac{h^4}{\mu k}\left(\frac{dp}{dx}\right)^2\left[\xi^2+\frac{1}{a^2}+\frac{1}{a^2}\frac{\cosh 2a\xi}{\cosh^2 a}-\frac{2\xi}{a}\frac{\sinh a\xi}{\cosh a}-\frac{2}{a^2}\frac{\cosh a\xi}{\cosh a}\right] \tag{3.12.7}$$

The solution of this equation, satisfying $\theta = 0$ at $\xi = \pm 1$, is

$$\begin{aligned}\Theta(\xi)&=\theta(\xi)/\left[\frac{h^4}{12\mu k}\left(\frac{dp}{dx}\right)^2\right]\\&=1-\xi^4+\frac{6}{a^2}(1-\xi^2)+\frac{3}{a^4\cosh^2 a}(\cosh 2a-\cosh 2a\xi)\\&\quad-\frac{24}{a^3\cosh a}(\sinh a-\xi\sinh a\xi)+\frac{24}{a^4}\left(1-\frac{\cosh a\xi}{\cosh a}\right)\end{aligned} \tag{3.12.8}$$

This equation gives the steady-state temperature distribution in Poiseuille flow between two plates which are maintained at a constant temperature T_0. The temperature distribution, shown graphically in Fig. 3.12.1, reduces to

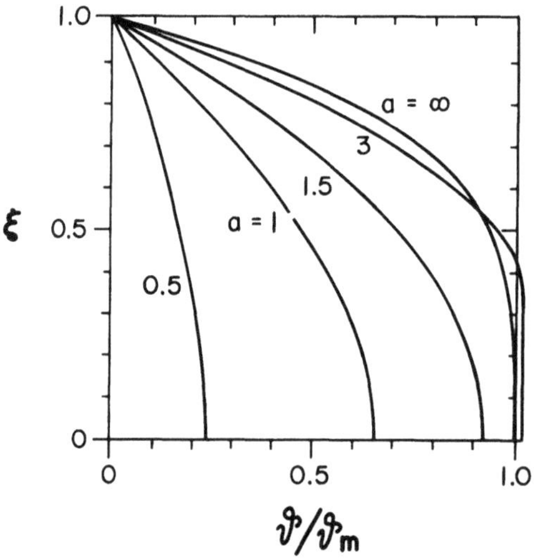

Fig. 3.12.1 Temperature distribution as a function of the parameter a. *Adapted from Ref. [5].*

the well-known nonpolar expression $\Theta(\xi) = 1 - \xi^4$ when $a \rightarrow \infty$. Also, $\Theta(\xi) \equiv 0$ as $a \rightarrow 0$, as expected, for as $a \rightarrow 0$ there is no flow and, therefore, under steady-state conditions the temperature will be uniform throughout the flow. The effects of couple stresses are quite significant for small values of a. For a certain range of values of a, the maximum temperature exceeds that in the nonpolar case.

The maximum temperature Θ_m occurs at $\xi = 0$. Also, in the nonpolar case is $\Theta_{m_o} = 1$. Therefore

$$\frac{\Theta_m}{\Theta_{m_o}} = \Theta_m = 1 + \frac{6}{a^2} + \frac{6}{a^4}\tanh^2 a - \frac{24}{a^3}\tanh a + \frac{24}{a^4}\left[1 - \frac{1}{\cosh a}\right] \tag{3.12.9}$$

The limiting values of Θ are given by $\lim_{a \rightarrow \infty} \Theta = 1$ and $\lim_{a \rightarrow 0} \Theta_m = 0$. The variation of Θ_m versus a is shown in Fig. 3.12.2. The range of values of a for which the maximum temperature is greater than that in the nonpolar case.

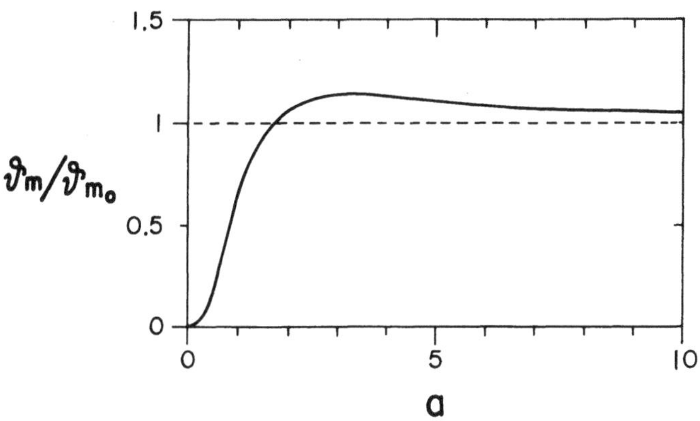

Fig. 3.12.2 Maximum temperature as a function of the parameter a. *Adapted from Ref. [5].*

3.13 Concluding Remarks

The theory presented in this chapter is incomplete in the sense that the skew-symmetric part of the stress tensor and the trace of the couple stress tensor are not determined by the constitutive equations. The equations of motion determine T_{ij}^A in terms of $m_{,s}$ as shown in Eq. (3.3.4). The trace, m, of the couple stress tensor is determined by the boundary conditions. However, the most objectionable part of the model is that the stress depends on the body moment, as can be inferred from Eq. (3.3.9). This difficulty will not arise in the absence of body moments.

The main feature of the theory of couple stresses, discussed in this chapter, is the prediction of a size-dependent effect, which is not present in the nonpolar theory. Therefore the first objective would be to look for a size-dependent effect, experimentally. In the nonpolar case, for plane Poiseuille flow, Eq. (3.5.21) gives $\mu = -\frac{2}{3}(h^3/Q_0)(dp/dx)$, irrespective of the magnitudes of h, Q_0 and dp/dx. However, when couple stresses are present, $-\frac{2}{3}(h^3/Q)(dp/dx)$ will not be a constant. In fact as h decreases, the value of this expression is expected to increase, especially for very small h. This length dependence can also be expected in other geometries. In terms of the nonpolar theory, this effect is equivalent to an apparent increase in the viscosity of the fluid.

Two types of boundary conditions have been discussed. In principle, the constants μ, η and η' may be determined for boundary conditions A as follows: Plane Couette flow determines μ as the effects of η and η' are absent. Plane Poiseuille flow is affected only by μ and η and not by η'. Once μ has been determined, this flow may then be used for determining η. Finally, since pipe Poiseuille flow is affected by μ, η and η', it may be used for determining η'.

Actually both μ and η can be determined from plane Poiseuille flow itself. For, as $a \rightarrow \infty$, that is as h is increased, $Q \rightarrow Q_0$. Thus Eq. (3.5.22) gives $\mu = -\lim_{h \rightarrow \infty} [\frac{2}{3}(h^3/Q)(dp/dx)]$. Experiments are performed with varying h, Q and dp/dx. A stabilization of the value of μ, calculated from the above expression, indicates that h is sufficiently large for determining the value of μ.

The validity of the boundary conditions can, in principle, again be checked experimentally. For boundary conditions A, the effects of couple stresses are absent in plane and cylindrical Couette flow. On the other hand, if boundary conditions B are used, then these flows are affected by couple stresses in that a size-dependent effect is predicted.

The effects of couple stresses are magnified in hydromagnetic channel flows. In principle, such flows could therefore be used to experimentally determine the effects of couple stresses.

Note that this theory is a special case of the more general theory of Eringen, discussed in Chapter 6, in which microstructure is taken into account.

3.14 References

The bulk of the results in this chapter are based on the material contained in Refs. [1] through [8]. Four related theories, which have not been considered in this book, are discussed in Refs. [9] to [12]. The remaining references contain additional solutions and applications of the theory discussed in this chapter.

1. Mindlin, R. D., and Tiersten, H. F. (1962). Effects of Couple Stresses in Linear Elasticity, *Arch. Ration. Mech. Anal.* **11**, 415-448.
2. Mindlin, R. D. (1963). Influence of Couple Stresses in Linear Elasticity, *Exp. Mech.* **20**, 1-7.
3. Stokes, V. K. (1966). Couple Stresses in Fluids, *Phys. Fluids* **9**, 1709-1715.
4. Stokes, V. K. (1968). Effects of Couple Stresses in Fluids on Hydromagnetic Channel Flows, *Phys. Fluids* **11**, 1131-1133.
5. Stokes, V. K. (1969). On Some Effects of Couple Stresses in Fluids on Heat Transfer, *J. Heat Transfer* **91**, 182-184.
6. Usmani, S. A. (1969). Some Time Dependent Flows with Couple Stresses, Master's Thesis, Department of Mechanical Engineering, Indian Institute of Technology, Kanpur, India.
7. Stokes, V. K. (1971). Effects of Couple Stresses in Fluids on the Creeping Flow Past a Sphere, *Phys. Fluids* **14**, 1580-1582.
8. Jain, J. K., and Stokes, V. K. (1972). Effects of Couple Stresses on the Stability of Plane Poiseuille Flow, *Phys. Fluids* **15**, 977-980.
9. Condiff, D. W., and Dahler, J. S. (1964) Fluid Mechanical Aspects of Antisymmetric Stress, *Phys. Fluids* **7**, 842-854.
10. Aero, E. L., Bulygin, A. N., and Kuvshinskii, E. V. (1965). Asymmetric Hydromechanics, *J. App. Math. Mech. (USSR)* **29**, 333-346.
11. Bleustein, J. L., and Green, A. E. (1967). Dipolar Fluids, *Int. J. Eng. Sci.* **5**, 323-340.
12. Cowin, S. C. (1968). Polar Fluids, *Phys. Fluids* **11**, 1919-1927.
13. Ariman, T., and Cakmak, A. S. (1967). Couple Stresses in Fluids, *Phys. Fluids* **10**, 2497-2499
14. Valanis, K. C., and Sun, C. T. (1969). Poiseuille Flow of a Fluid with Couple Stresses with Applications to Blood Flow, *Biorheology* **6**, 85-97.
15. Soundalgekar, V. M. (1971). Effects of Couple Stresses in Fluids on Dispersion of a Solute in a Channel Flows, *Phys. Fluids* **14**, 19-20.
16. Soundalgekar, V. M., and Aranake, R. N. (1974). Effects of Couple Stresses on the Oscillatory Flow Past an Infinite Plate with Constant Suction, *Meccanica* **9**, 194-198.
17. Annapurna, N., and Ramanaiah, G. (1976). Effect of Couple Stresses on the Unsteady Drainage of a Micro-Polar Fluid on a Flat Surface, *Jpn. J. Appl. Phys.* **15**, 2441-2444.

18. Chaturani, P. (1976). Some Comments on Poiseuille Flow of a Liquid with Couple Stresses with Applications to Blood Flow, *Biorheology* **13**, 133-136.

19. Soundalgekar, V. M. (1976). Effects of Couple Stresses in Fluids on the Dispersion of a Soluble Matter in a Channel Flow with Homogeneous and Heterogeneous Reactions, *Int. J. Heat Mass Transfer* **18**, 527-530.

20. Chaturani, P., and Upadhya, V. S. (1977). Gravity Flow of a Fluid with Couple Stress Along an Inclined Plane, *Biorheology* **14**, 237-246.

21. Soundalgekar, V. M., and Aranake, R. N. (1977). Effects of Couple Stresses in MHD Channel Flow, *Nucl. Eng. Des.* **44**, 301-308.

22. Chaturani, P. (1978). Viscosity of Poiseuille Flow of a Couple Stress Fluid with Applications to Blood Blow, *Biorheology* **15**, 119-128.

23. Chaturani, P., and Upadhya, V. S. (1978). Pulsatile Flow of a Couple Stress Fluid Through Circular Tubes with Applications to Blood Flow, *Biorheology* **15**, 193-201.

24. Rammkissoon, H. (1978). Drag in Couple Stress Fluids, *Z. Angew. Math. Phys.* **29**, 341-346.

25. Ramanaiah, G., and Sarkar, P. (1979). Slider Bearings Lubricated by Fluids with Couple Stress, *Wear* **52**, 27-36.

Chapter 4

Anisotropic Fluids

4.1 Introduction

This chapter considers some effects that occur when microstructure of a fluid is taken into account. Microstructure has been successfully accounted for by J. L. Ericksen in his theory of anisotropic fluids. Such theories are an outcome of an effort to explain the behavior of liquid crystals.

In these theories, it is assumed that the velocity field is not sufficient for describing the kinematics of flow as, in addition to this field, each material particle may be spinning independently of its gross motion, thereby contributing an additional term to the angular momentum. A visualization of this is not difficult if a material is thought of as being an assemblage of independent particles that are at a distance from each other. These particles can spin about an axis independent of their motion in space. The difficulty is in extending this concept to a continuum.

In order to emphasize the effects of the microstructure of a fluid, this chapter assumes that couple stresses are absent.

The results described in this chapter are based on the pioneering work of J. L. Ericksen listed in the references at the end of this chapter.

4.2 Balance Laws

With $\mathbf{p} = \mathbf{v}$, $\mathbf{h} = \mathbf{x} \times \mathbf{p} + \boldsymbol{\sigma}$ and $e = \epsilon + \frac{1}{2} v_r v_r + k_s$, if couple stresses are assumed to be zero, then from the results of Sections (2.3), (2.6) and (2.8) the basic laws for the balance of mass, linear momentum, angular momentum, and energy are given, respectively, by the equations

$$\dot{\rho} + \rho v_{r,r} = 0 \tag{4.2.1}$$

$$\rho a_i = t_{ri,r} + \rho f_i \tag{4.2.2}$$

$$\rho \dot{\sigma}_i = e_{ijk} t_{jk} + \rho l_i \tag{4.2.3}$$

$$\rho \dot{\epsilon} = \rho(\omega_r \dot{\sigma}_r - \dot{k}_s) + t_{rs} d_{rs} - q_{r,r} + \rho h \tag{4.2.4}$$

where σ_i is the spin, or intrinsic angular momentum per unit mass and k_s is the kinetic energy of spin per unit mass. Both σ_i and k_s are introduced into the analysis by a consideration of microstructure. Suitable expressions must be found for σ_i and k_s so that Eqs. (4.2.3) and (4.2.4) can be used for studying the effects of microstructure.

A heuristic guide is provided by considering the expressions for linear and angular momenta for particles having structure, such as, for example, dumbbell-shaped particles.

4.3 Microstructure of a Dumbbell-Shaped Particle

Consider a dumbbell-shaped "particle," consisting of two masses m_1 and m_2 that are separated by a small distance, as shown in Fig. 4.3.1. If the

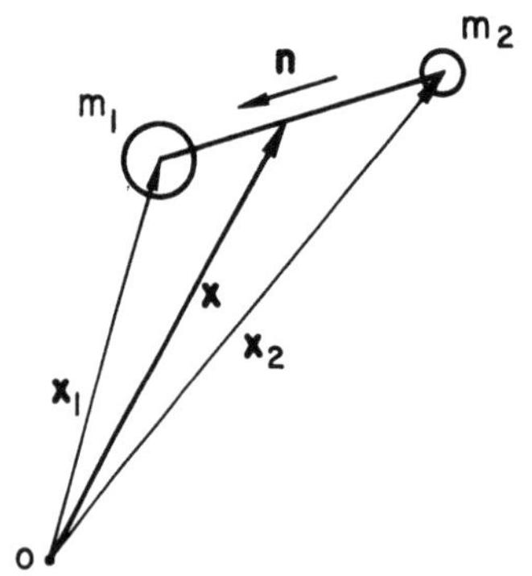

Fig. 4.3.1 Geometry of a dumbbell-shaped particle.

masses are at $\mathbf{x}_1$ and $\mathbf{x}_2$, respectively, then the linear momentum of the particle, $\mathbf{P}$, is given by $m_1\dot{x}_1 + m_2\dot{x}_2 = \dfrac{D}{Dt}(m_1\mathbf{x}_1 + m_2\mathbf{x}_2) = m\dot{\mathbf{x}}$, that is

$$\mathbf{P} = m\dot{\mathbf{x}} \tag{4.3.1}$$

where $m = m_1 + m_2$ is the mass of the particle and $\mathbf{x}$ its center of mass, so that

$$m\mathbf{x} = m_1\mathbf{x}_1 + m_2\mathbf{x}_2 \quad , \quad m = m_1 + m_2 \tag{4.3.2}$$

Now define a vector, along the line joining m_2 to m_1, by

$$m\mathbf{n} = (m_1 m_2)^{1/2}(\mathbf{x}_1 - \mathbf{x}_2) \tag{4.3.3}$$

Then, Eqs. (4.3.2) and (4.3.3) may be solved to give

$$\mathbf{x}_1 = \mathbf{x} + \frac{m_2}{(m_1 m_2)^{1/2}}\mathbf{n} \quad , \quad \mathbf{x}_2 = \mathbf{x} - \frac{m_1}{(m_1 m_2)^{1/2}}\mathbf{n} \tag{4.3.4}$$

The angular momentum $\mathbf{H}$ of the particle is given by $\mathbf{x}_1\times(m_1\dot{\mathbf{x}}_1) + \mathbf{x}_2\times(m_2\dot{\mathbf{x}}_2)$, which, upon using the expressions for $\mathbf{x}_1$ and $\mathbf{x}_2$ given in Eq. (4.3.4), simplifies to $\mathbf{x}\times(m_1\dot{\mathbf{x}} + m_2\dot{\mathbf{x}}) + (m_1 + m_2)\mathbf{n}\times\dot{\mathbf{n}}$, so that

$$\mathbf{H} = m(\mathbf{x}\times\dot{\mathbf{x}} + \mathbf{n}\times\dot{\mathbf{n}}) \tag{4.3.5}$$

The kinetic energy K of the system is given by $\frac{1}{2}(m_1\dot{\mathbf{x}}_1\cdot\dot{\mathbf{x}}_1 + m_2\dot{\mathbf{x}}_2\cdot\dot{\mathbf{x}}_2)$ which, upon using Eq. (4.3.4) to eliminate $\mathbf{x}_1$ and $\mathbf{x}_2$, gives

$$K = \frac{1}{2}m(\dot{\mathbf{x}}\cdot\dot{\mathbf{x}} + \dot{\mathbf{n}}\cdot\dot{\mathbf{n}}) \tag{4.3.6}$$

Thus, for a dumbbell-shaped particle of mass m, the linear momentum, the angular momentum, and the kinetic energy are given by Eqs. (4.3.1), (4.3.5) and (4.3.6), respectively, so that the expressions for these quantities per unit mass of the particle would be $\dot{\mathbf{x}}$, $\mathbf{x}\times\dot{\mathbf{x}} + \mathbf{n}\times\dot{\mathbf{n}}$, and $\frac{1}{2}(\dot{\mathbf{x}}\cdot\dot{\mathbf{x}} + \dot{\mathbf{n}}\cdot\dot{\mathbf{n}})$, respectively.

At this stage, if a fluid is considered to be an aggregation of small particles, each of which has an associated preferred direction $\mathbf{n}$, then an appropriate generalization for a fluid with microstructure is given by

$$\mathbf{p} = \dot{\mathbf{x}} \tag{4.3.7}$$

$$\mathbf{h} = \mathbf{x}\times\dot{\mathbf{x}} + \mathbf{n}\times\dot{\mathbf{n}} \tag{4.3.8}$$

$$k = \frac{1}{2}(\dot{\mathbf{x}}\cdot\dot{\mathbf{x}} + \dot{\mathbf{n}}\cdot\dot{\mathbf{n}}) \tag{4.3.9}$$

Then, from the definitions of $\boldsymbol{\sigma}$ and k_s,

$$\boldsymbol{\sigma} = \mathbf{n}\times\dot{\mathbf{n}} \quad , \quad k_s = \frac{1}{2}\dot{\mathbf{n}}\cdot\dot{\mathbf{n}} \tag{4.3.10}$$

4.4 Field Equations

An anisotropic fluid is defined as one in which each particle has a preferred direction. For describing its motion, in addition to the velocity $\mathbf{v}$ of the particle, an additional vector $\mathbf{n}$ that describes its orientation, has to be considered. Following the discussion in Section (4.3), the densities $\mathbf{p}$, $\boldsymbol{\sigma}$ and k_s of the linear momentum, the spin momentum, and the kinetic energy of spin will be assumed to be given, respectively, by Eqs. (4.3.7), (4.3.8) and (4.3.9), so that

$$p_i = v_i \tag{4.4.1}$$

$$\sigma_i = e_{ijk} n_j \dot{n}_k \tag{4.4.2}$$

$$k_s = \frac{1}{2} \dot{n}_r \dot{n}_r \tag{4.4.3}$$

and $\dot{\mathbf{p}} = \mathbf{a}$, $\dot{\boldsymbol{\sigma}} = \mathbf{n} \times \ddot{\mathbf{n}}$, $\dot{k}_s = \dot{\mathbf{n}} \cdot \ddot{\mathbf{n}}$.

Using Eq. (4.4.2), Eq. (4.4.3) becomes

$$\rho e_{ijk} n_j \ddot{n}_k = e_{ijk} t_{jk} + \rho l_i \tag{4.4.4}$$

which can be inverted to give

$$t_{ij}^A = t_{[ij]} = \frac{1}{2} \rho [n_i \ddot{n}_j - n_j \ddot{n}_i] - \frac{1}{2} \rho e_{ijk} l_k \tag{4.4.5}$$

Thus, in general, the stress tensor will not be symmetric, even if the body moment $\mathbf{l}$ is zero.

From Eqs. (4.4.2) and (4.4.3)

$$\rho (\dot{k}_s - \omega_r \dot{\sigma}_r) = \rho (\dot{n}_r + w_{rs} n_s) \ddot{n}_r \tag{4.4.6}$$

where use has been made of $w_{rs} = e_{rst} \omega_t$. By substituting from this equation in Eq. (4.2.4), the energy equation can be written as

$$\rho \dot{\epsilon} = -\rho (\dot{n}_r + w_{rs} n_s) \ddot{n}_r + t_{rs} d_{rs} - q_{r,r} + \rho h \tag{4.4.7}$$

Since $\mathbf{n}$ is associated with a particle, it is an objective vector. That is, under a time-dependent coordinate transformation given by $\bar{x}_i = Q_{ir}(t) x_r + c_i(t)$, $Q_{ir} Q_{jr} = \delta_{ij}$, its components transform as $\bar{n}_i = Q_{ir} n_r$. A time differentiation of this equation gives

$$\dot{\bar{n}}_i = Q_{ir} \dot{n}_r + \dot{Q}_{ir} n_r \tag{4.4.8}$$

which shows that $\dot{\mathbf{n}}$ is not objective. In its place Ericksen introduced the objective vector

$$\hat{n}_i = \dot{n}_i + w_{ir} n_r \tag{4.4.9}$$

That $\hat{n}_i$ is objective, can be shown as follows: by definition

$$\hat{\bar{n}}_i = \dot{\bar{n}}_i + \bar{w}_{ij} \bar{n}_j$$

so that a use of Eqs. (4.4.8) and (1.7.6) gives

$$\hat{\bar{n}}_i = (Q_{ij} \dot{n}_j + \dot{Q}_{ij} n_j) + (Q_{ir} Q_{js} w_{rs} + Q_{ir} \dot{Q}_{jr}) Q_{jt} n_t$$

$$= Q_{ij} \dot{n}_j + \dot{Q}_{ij} n_j + Q_{ir} Q_{js} Q_{jt} w_{rs} n_t + Q_{ir} Q_{jt} \dot{Q}_{jr} n_t$$

On the right-hand side of this equation, the third term reduces to $Q_{ir} w_{rs} n_s$ and, since $Q_{jt} Q_{jr} = \delta_{tr}$ implies that $Q_{jt} \dot{Q}_{jr} = -\dot{Q}_{jt} Q_{jr}$, the last term is given by $-Q_{ir} \dot{Q}_{jt} Q_{jr} n_t = -\dot{Q}_{it} n_t$, so that

$$\bar{\hat{n}}_i = Q_{ir}(\dot{n}_r + w_{rs} n_s) = Q_{ir} \hat{n}_r$$

Use of Eq. (4.4.9) reduces Eq. (4.4.6) to

$$\rho(\dot{k}_s - \omega_{,r} \dot{\sigma}_{,r}) = \rho \hat{n}_r \ddot{n}_r \tag{4.4.10}$$

Substituting from Eq. (4.4.9) in Eq. (4.4.7), the energy equation can be written as

$$\rho \dot{\epsilon} = t_{rs} d_{rs} - \rho \hat{n}_r \ddot{n}_i - q_{r,r} + \rho h \tag{4.4.11}$$

Since $\ddot{n}_r$ appears in combination with the objective kinematic measure $\hat{n}_r$, $\ddot{n}_r$ may be regarded as a quantity that measures an interaction for a particular material, and is therefore to be determined by a constitutive equation. Following the work of Oseen [Ref. 11], Ericksen introduced the expression

$$\rho \ddot{n}_i = g_i \tag{4.4.12}$$

where g_i is a quantity that is to be determined by a constitutive equation.

The governing equations may therefore be summarized as

$$\dot{\rho} + \rho v_{r,r} = 0 \tag{4.4.13}$$

$$\rho \dot{v}_i = t_{ri,r} + \rho f_i \tag{4.4.14}$$

$$t_{ij}^A = \frac{1}{2}(n_i g_j - n_j g_i) - \frac{1}{2} \rho e_{ijk} l_k \tag{4.4.15}$$

$$\rho \dot{\epsilon} = t_{rs} d_{rs} - g_i \hat{n}_i - q_{r,r} + \rho h \tag{4.4.16}$$

where $\rho \ddot{n}_i = g_i$.

The next step is to consider suitable constitutive equations for the fields t_{ij}, g_i and q_i. A linear model proposed by Ericksen is discussed in the next section.

4.5 Constitutive Equations

If it is assumed that t_{ij}, g_i and q_i, at each point P at time t, are all functions of the variables ρ, θ, n_i, $\dot{n}_i$, $v_{j,i}$ and $\theta_{,i}$ evaluated at P at time t, then an application of the principle of material objectivity will imply that the proper objective variables are

$$\rho \ , \ \theta \ , \ n_i \ , \ \hat{n}_i \ , \ d_{ij} \ , \ \theta_{,i} \tag{4.5.2}$$

In this chapter, only the special case in which the dependent variables are linearly dependent on the independent variables $\hat{n}_i$, d_{ij} and $\theta_{,i}$ will be considered, so that

$$\left.\begin{aligned} t_{ij} &= A_{ij}^0 + A_{ijk}^1 \hat{n}_k + A_{ijkm}^2 d_{km} + A_{ijk}^3 \theta_{,k} \\ q_i &= B_i^0 + B_{ij}^1 \hat{n}_j + B_{ijk}^2 d_{jk} + B_{ij}^3 \theta_{,j} \\ g_i &= C_i^0 + C_{ij}^1 \hat{n}_j + C_{ijk}^2 d_{jk} + C_{ij}^3 \theta_{,j} \end{aligned}\right\} \tag{4.5.3}$$

where the A's, B's and C's are functions of ρ, θ and n_i.

Additionally, the constitutive equations given in Eqs. (4.5.3) will be assumed to be invariant under reflections through all planes containing $\mathbf{n}$. This assumption implies that the A's, B's and C's are transversely isotropic tensors with respect to the direction $\mathbf{n}$. Such tensors have been studied by Smith and Rivlin [Ref. 15], who show that these terms are expressible as linear combinations of the outer products formed from the tensors n_i and $\delta_{ij} - n_i n_j$, or, equivalently

$$n_i, \ \delta_{ij} \tag{4.5.4}$$

For the situation under discussion, the scalar coefficients in these combinations can then be shown to reduce to functions of

$$\rho \quad , \quad \theta \quad , \quad n^2 = n_r n_r \tag{4.5.5}$$

Finally, the dependent variables will be assumed to transform as

$$t_{ij} \rightarrow t_{ij} \quad , \quad q_i \rightarrow q_i \quad , \quad g_i \rightarrow -g_i \tag{4.5.6}$$

when $\mathbf{n}$ and $\hat{\mathbf{n}}$ are mapped as

$$n_i \rightarrow -n_i \quad , \quad \hat{n}_i \rightarrow -\hat{n}_i \tag{4.5.7}$$

which is equivalent to assuming that $\mathbf{n}$ and $-\mathbf{n}$ are physically indistinguishable. For example, in the model discussed in Section (4.3), if $m_1 = m_2$, then $\mathbf{n}$ and $-\mathbf{n}$ would be physically indistinguishable. From Eqs. (4.5.3), (4.5.6) and (4.5.7), and the results summarized in Eq. (4.5.4), the constitutive equations listed in Eq. (4.5.3) reduce to

$$\begin{aligned} t_{ij} &= (\alpha_0 + \alpha_1 d_{rr} + \alpha_2 d_{rs} n_r n_s + \alpha_3 \hat{n}_r n_r)\delta_{ij} \\ &\quad + (\alpha_4 + \alpha_5 d_{rr} + \alpha_6 d_{rs} n_r n_s + \alpha_7 \hat{n}_r n_r) n_i n_j \\ &\quad + \alpha_8 d_{ij} + \alpha_9 d_{ir} n_r n_j + \alpha_{10} d_{jr} n_r n_i \end{aligned} \tag{4.5.8}$$

$$+ \alpha_{11} n_i \hat{n}_j + \alpha_{12} n_j \hat{n}_i$$

$$q_i = \beta_0 \theta_{,i} + \beta_1 n_r \theta_{,r} n_i \tag{4.5.9}$$

$$g_i = (\gamma_0 + \gamma_1 d_{rr} + \gamma_2 d_{rs} n_r n_s + \gamma_3 \hat{n}_r n_r) n_i$$

$$+ \gamma_4 d_{ir} n_r + \gamma_5 \hat{n}_i \tag{4.5.10}$$

where the α's, β's and γ's are functions of the variables listed in Eqs. (4.5.5). In order for the number of equations to equal the number of unknowns, the equation $t_{[ij]} = \frac{1}{2}(n_i g_j - n_j g_i) - \frac{1}{2}\rho e_{ijk} l_k$ must hold as an identity which, in the absence of body moments, requires that

$$\gamma_4 = \alpha_{10} - \alpha_9 \quad , \quad \gamma_5 = \alpha_{11} - \alpha_{12} \tag{4.5.11}$$

Equation (4.4.15) shows one of the weak points of this theory in that the stress, through $t_{[ij]}$, depends on the body moment. In his original papers Ericksen did not allow for body moments, so that Eq. (4.5.11) could be imposed as a requirement on the constitutive equations. A similar weakness was pointed out for the theory for couple stresses in fluids discussed in the last chapter, where Eq. (3.3.9) shows a dependence of the stress t_{ij} on the body moment l_i.

4.6 Implications of the Second Law of Thermodynamics

The internal energy density ϵ, which occurs in the energy equation given by Eq. (4.4.11), will be assumed to have an equation of state of the form

$$\epsilon = \epsilon(\eta, \rho, n_i) \tag{4.6.1}$$

where η is the entropy per unit mass. The invariance requirement imposed by the principle of material objectivity then implies that ϵ must have the form

$$\epsilon = \epsilon(\eta, \rho, n^2) \tag{4.6.2}$$

so that, in Eq. (2.9.9), $\nu_1 = \rho$, $\nu_2 = n^2$ and all the other ν_α's are zero.

Using Eqs. (4.4.10) and (2.9.12), the rate of entropy production per unit mass, γ, is given by

$$\rho\theta\gamma = t_{(rs)} d_{rs} - \rho \hat{n}_r \dot{n}_r - \rho \left[\frac{\partial \epsilon}{\partial \rho} \dot{\rho} + \frac{\partial \epsilon}{\partial n^2} \dot{\overline{n^2}} \right] - q_r (\ln \theta)_{,r} \tag{4.6.3}$$

From the continuity equation $\dot{\rho} = -\rho v_{r,r} = -\rho\delta_{rs}d_{rs}$. Also $\hat{n}_r = \dot{n}_r + w_{rs}n_s$ results in $\hat{n}_r n_r = n_r \dot{n}_r$, so that $\dot{\overline{n^2}} = 2n_r\dot{n}_r = 2n_r\hat{n}_r$. Equation (4.6.4) can then be written as

$$\rho\theta\gamma = (t_{(rs)} + \rho^2 \frac{\partial\epsilon}{\partial\rho}\delta_{rs})d_{rs} - (g_r + 2\rho\frac{\partial\epsilon}{\partial n^2}n_r)\hat{n}_r - q_r(\ln\theta)_{,r} \tag{4.6.4}$$

since $\rho\ddot{n}_r = g_r$.

If, instead of ϵ, the free energy

$$\psi = \epsilon - \theta\eta = \psi(\rho, n^2, \theta) \tag{4.6.5}$$

is used, then, by using the results

$$\frac{\partial\epsilon}{\partial\rho}\Big|_{\eta,n^2} = \frac{\partial\psi}{\partial\rho}\Big|_{\theta,n^2}\ , \quad \frac{\partial\epsilon}{\partial n^2}\Big|_{\eta,\rho} = \frac{\partial\psi}{\partial n^2}\Big|_{\theta,\rho}\ , \quad \eta = -\frac{\partial\psi}{\partial\theta} \tag{4.6.6}$$

Eq. (4.6.4) can be written as

$$\rho\theta\gamma = (t_{(rs)} + \rho^2 \frac{\partial\psi}{\partial\rho}\delta_{rs})d_{rs} - (g_r + 2\rho\frac{\partial\psi}{\partial n^2}n_r)\hat{n}_r - q_r(\ln\theta)_{,r} \tag{4.6.7}$$

The Clausius-Duhem inequality is then of the form

$$\gamma \geqslant 0 \tag{4.6.8}$$

where γ is given by either one of Eqs. (4.6.4) and (4.6.7). This inequality can be made more explicit by using the constitutive equations, given in Eqs. (3.4.8) to (3.4.10), in which case it will impose restrictions on the coefficients occurring in these equations.

4.7 *Incompressible Fluids*

The equations that describe the flow may now be summarized as

$$\dot{\rho} + \rho v_{r,r} = 0$$

$$\rho\dot{v}_i = t_{ri,r} + \rho f_i$$

$$\rho\ddot{n}_i = g_i$$

$$\rho\theta\dot{\eta} = (t_{(rs)} + \rho^2 \frac{\partial\psi}{\partial\rho}\delta_{rs})d_{rs} - (g_r + 2\rho\frac{\partial\psi}{\partial n^2}n_r)\hat{n}_r - q_{r,r} + \rho h$$

where use has been made of Eq. (2.9.5). This is a set of eight equations for the eight unknowns: ρ, θ, v_i and n_i, with the understanding that the body force is specified and that t_{ij}, q_i and g_i are given by the constitutive equations, given by Eqs. (4.5.8) to (4.5.10), and that $\psi = \epsilon - \theta\eta$.

In general, it is consistent with these equations to take $n_i = 0$. The governing equations then reduce to those used for isotropic fluids, which will eventually result in the Navier-Stokes equations.

A further assumption could be that the "molecular inertia," measured by $\rho \ddot{n}_i$, is negligible, so that the equation $\rho \ddot{n}_i = g_i$ can be replaced by $g_i = 0$. This can then be solved for $\hat{n}_i$ to give an equation of the form

$$\hat{n}_i = h_i(\rho, \theta, n_i, d_{ij}) \tag{4.7.1}$$

Finally, this relation could be used to eliminate $\hat{n}_i$ from the expression for the stress.

Now consider the isothermal flow of an incompressible fluid. Then, the equations describing thermal interactions, such as Eqs. (4.5.9) and (4.6.8), need not be considered. Because of the incompressibility condition, terms proportional to d_{rr} will drop out. Assuming that the "molecular inertia" term, g_i, is zero, the constitutive equations given in Eqs. (4.5.8) and (4.5.10) reduce to

$$\begin{aligned} t_{ij} = {} & (\alpha_0 + \alpha_2 d_{rs} n_r n_s + \alpha_3 \hat{n}_r n_r)\delta_{ij} \\ & + (\alpha_4 + \alpha_6 d_{rs} n_r n_s + \alpha_7 \hat{n}_r n_r) n_i n_j \\ & + \alpha_8 d_{ij} + \alpha_9 d_{ir} n_r n_j + \alpha_{10} d_{jr} n_r n_i \\ & + \alpha_{11} n_i \hat{n}_j + \alpha_{12} n_j \hat{n}_i \end{aligned} \tag{4.7.2}$$

$$0 = (\gamma_0 + \gamma_2 d_{rs} n_r n_s + \gamma_3 \hat{n}_r n_r) n_i + \gamma_r d_{ir} n_r + \gamma_5 \hat{n}_i \tag{4.7.3}$$

where the α's and γ's are now only functions of n^2, as ρ and θ are fixed.

An inner multiplication of Eq. (4.7.3) by n_i can be solved to give

$$\hat{n}_r n_r = -\frac{1}{n^2\gamma_3 + \gamma_5}\left[n^2\gamma_0 + (n^2\gamma_2 + \gamma_4) d_{rs} n_r n_s\right] \tag{4.7.4}$$

which, upon substitution in Eq. (4.7.3), gives

$$\hat{n}_i = (\mu_1 + \mu_2 d_{rs} n_r n_s) n_i + \mu_3 d_{ir} n_r \tag{4.7.5}$$

where μ_1, μ_2 and μ_3 are functions of the γ's and hence depend on n^2.

Now, in the absence of body moments if $g_i = 0$ as assumed, then Eq. (4.4.15) implies that $t_{[ij]} = 0$, so that t_{ij} must be symmetric. Symmetry of **T** requires that $\alpha_9 = \alpha_{10}$ and $\alpha_{11} = \alpha_{12}$ in Eq. (4.7.2). If the coefficient of δ_{ij} in Eq. (4.7.2) is assumed to be an arbitrary scalar, then a substitution for $\hat{n}_i$ from Eq. (4.3.5) in Eq. (4.7.2) reduces the expression for t_{ij} to

$$t_{ij} = -p\delta_{ij} + (\lambda_1 + \lambda_2 d_{rs} n_r n_s) n_i n_j + 2\lambda_3 d_{ij} + 2\lambda_4 (d_{ir} n_r n_j + d_{jr} n_r n_i) \tag{4.7.6}$$

Thus, the assumptions: (i) that only isothermal flows of an incompressible fluid are being considered, (ii) that the coefficient of δ_{ij} in the expression for t_{ij} can be replaced by an arbitrary scalar, and (iii) that the "molecular inertia" g_i is negligible, result in a reduction of the constitutive equations to the form

$$t_{ij} = -p\delta_{ij} + (\lambda_1 + \lambda_2 d_{rs} n_r n_s) n_i n_j + 2\lambda_3 d_{ij} + 2\lambda_4 (d_{ir} n_r n_j + d_{jr} n_r n_i) \tag{4.7.7}$$

$$\hat{n}_i = (\mu_1 + \mu_2 d_{rs} n_r n_s) n_i + \mu_3 d_{ir} n_r \tag{4.7.8}$$

where the λ's and μ's are, for fixed ρ and θ, only functions of n^2.

A further assumption could be that the magnitude of $\mathbf{n}$ is constant, as would be true for rigid "inelastic" molecules. For example, this assumption would imply that the masses m_1 and m_2, in the dumbbell model discussed in Section (4.3), always remain at the same distance from each other. A variable $|\mathbf{n}|$ would have implied that this distance can change.

If $n^2 = n_i n_i$ is assumed to be a constant then, since $g_i = 0$, $\hat{n}_i n_i = n_i \dot{n}_i = \frac{1}{2}\frac{d}{dt}(n^2) = 0$, so that use of Eq. (4.7.8) gives $n^2\mu_1 + (n^2\mu_2 + \mu_3) d_{rs} n_r n_s = 0$, which is only possible if

$$\mu_1 = 0 \quad , \quad n^2\mu_2 + \mu_3 = 0 \tag{4.7.9}$$

For the case in which n^2 is a constant, there is no loss of generality in taking this value to be unity, as its magnitude could be absorbed in the coefficients λ's and μ's. Thus, without any loss of generality, $\mathbf{n}$ will be assumed to be a unit vector so that the constitutive equations will be of the form

$$t_{ij} = -p\delta_{ij} + (\lambda_1 + \lambda_2 d_{rs} n_r n_s) n_i n_j + 2\lambda_3 d_{ij} + 2\lambda_4 (d_{ij} n_r n_j + d_{jr} n_r n_i) \tag{4.7.10}$$

$$\hat{n}_i = \mu_3 (d_{ir} n_r - d_{rs} n_r n_s n_i) \tag{4.7.11}$$

where the λ's and μ's are now constants, that are different from the constants in Eqs. (4.7.7) and (4.7.8), since the magnitude n of $\mathbf{n}$ has been absorbed in these constants. In fact, the constants λ_1, λ_2, λ_3, λ_4 and μ_3 in

Eqs. (4.7.10) and (4.7.11) are equal, respectively, to λ_1/n^2, λ_2/n^4, λ_3, λ_4/n^2 and μ_3/n in Eqs. (4.7.7) and (4.7.8).

4.8 Simple Shearing Motion

This section is based on the assumptions that the fluid is incompressible, that the "molecular inertia" can be neglected, and that **n** is a unit vector. If only isothermal flows are considered, then the relevant constitutive equations are given by Eqs. (4.7.10) and (4.7.10).

Now consider a flow of the type

$$v_1 = Kx_2 \quad , \quad v_2 = 0 \quad , \quad v_3 = 0 \tag{4.8.1}$$

where K is a constant. For this field, the only nonzero components of **D** and **W** are given by

$$d_{12} = \frac{1}{2}K \quad , \quad w_{12} = -\frac{1}{2}K \tag{4.8.2}$$

so that $d_{rs}n_r n_s = Kn_1 n_2$. From Eq. (4.7.10) the stresses are then given by

$$\left.\begin{aligned}
t_{11} &= -p + \lambda_1 n_1^2 + K(\lambda_2 n_1^2 + 2\lambda_4)n_1 n_2\\
t_{22} &= -p + \lambda_1 n_1^2 + K(\lambda_2 n_2^2 + 2\lambda_4)n_1 n_2\\
t_{33} &= -p + \lambda_1 n_3^2 + K\lambda_2 n_1 n_2 n_3^2\\
t_{23} &= \lambda_1 n_2 n_3 + K(\lambda_2 n_2^2 + \lambda_4)n_1 n_3\\
t_{31} &= \lambda_1 n_3 n_1 + K(\lambda_2 n_1^2 + \lambda_4)n_2 n_3\\
t_{12} &= \lambda_1 n_1 n_2 + K[\lambda_3 + \lambda_2 n_1^2 n_2^2 + \lambda_4(n_1^2 + n_2^2)]
\end{aligned}\right\} \tag{4.8.3}$$

These equations will satisfy the equations of steady motion, provided p and n_i are independent of the coordinates x_i.

Using the relation $\hat{n}_i = \dot{n}_i + w_{ir}n_r$ with Eq. (4.7.11), the components of $\dot{n}_i$ are given by

$$\left.\begin{aligned}
\dot{n}_1 &= \frac{1}{2}Kn_2(\mu_3 + 1 - 2\mu_3 n^2)\\
\dot{n}_2 &= \frac{1}{2}Kn_1(\mu_3 - 1 - 2\mu_3 n_2^2)\\
\dot{n}_3 &= -K\mu_3 n_1 n_2 n_3
\end{aligned}\right\} \tag{4.8.4}$$

From the first two of the above equations $n_1 dn_1/(\mu_3 + 1 - 2\mu_3 n_1^2) = n_2 dn_2/(\mu_3 - 1 - 2\mu_3 n_2^2)$, which can be integrated to give

$$an_1^2 + bn_2^2 = \frac{1}{2\mu_3}[(a+b)\mu_3 + a - b] \tag{4.8.5}$$

where a and b are arbitrary constants. This is an integral curve of Eq. (4.8.4), provided $\mu_3 \neq 0$, and can be written as

$$\frac{n_1^2}{\alpha^2 + \beta^2/r} + \frac{n_2^2}{r\alpha^2 + \beta^2} = 1 \tag{4.8.6}$$

where $\alpha^2 = (\mu_3 + 1)/2\mu_3$, $\beta^2 = (\mu_3 - 1)/2\mu_3$ and $r = a/b$. Furthermore, in terms of α^2 and β^2, the first two equations in Eq. (4.8.4) can be written as

$$\left.\begin{aligned} \dot{n}_1 &= K\mu_3 n_2(\alpha^2 - n_1^2) \\ \dot{n}_2 &= K\mu_3 n_1(\beta^2 - n_2^2) \end{aligned}\right\} \tag{4.8.7}$$

If $\mu_3 > 1$, α^2 and β^2 will be positive for all positive values of r. If $r = 0$, then Eq. (4.8.6) gives $n_2 = \pm\beta$, so that from Eq. (4.8.7) a, the sign of $\dot{n}_1$ is the same as that of n_2. If $r \to \infty$, then Eq. (4.8.6) gives $n_1 = \pm\alpha$ with the sign of $\dot{n}_2$ being the same as that of n_1. For $r > 0$, the integral curves will be ellipses. For negative values of r the curves will be hyperbolas. In each case, the integral curves will pass through the four points $(\alpha, \pm\beta)$ and $(-\alpha, \pm\beta)$.

For representative values of $r = a/b$, these integral curves for $\mu_3 > 1$, $\mu_3 = 1$ and $0 < \mu_3 < 1$, have been shown in Figs. 4.8.1 a, b and c, respectively, wherein the arrowheads indicate the direction of increasing time for $K > 0$. For $K < 0$, these arrowheads should be reversed. Corresponding plots for numerically equal negative values of μ_3 can be obtained by inter-

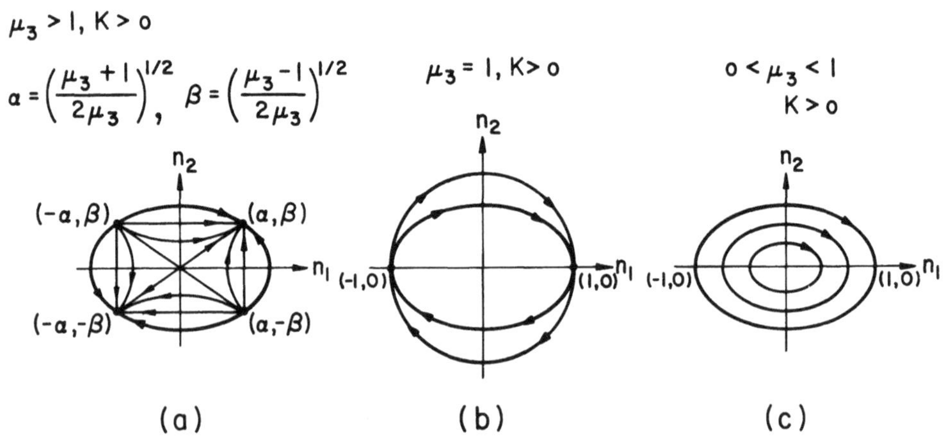

Fig. 4.8.1 Integral curves for simple shearing motion. *Adapted from Ref. [1].*

changing n_1 and n_2 and reversing the direction of the arrowheads. Since the solutions change rather drastically as $(\mu_3 - 1)$ changes sign, the cases $\mu_3 \geqslant 1$ and $0 < \mu_3 < 1$ are discussed separately.

1. Materials with $\mu_3 \geqslant 1$

For $\mu_3 > 1$, the above discussion and Fig. 4.8.1 a show that almost all solutions of Eq. (4.8.4) tend toward the constant solutions

$$\left.\begin{aligned} n_1 &= \epsilon\alpha = \epsilon[(\mu_3 + 1)/2\mu_3]^{1/2} \\ n_2 &= \epsilon\beta = \epsilon[(\mu_3 - 1)/2\mu_3]^{1/2} \\ n_3 &= 0 \end{aligned}\right\} \tag{4.8.8}$$

if $K > 0$, and

$$n_1 = \epsilon\alpha \quad , \quad n_2 = -\epsilon\beta \quad , \quad n_3 = 0 \tag{4.8.9}$$

if $K < 0$, where $\epsilon = \pm 1$.

There is thus a stable "steady state" solution. The exceptional solutions, which will be ignored for the present, are the unstable constant solutions $n_1 = n_2 = 0$, $n_3 = \pm 1$, and

$$\left.\begin{aligned} n_1 &= \epsilon\alpha \quad , \quad n_2 = -\epsilon\beta \quad , \quad \text{if} \quad K > 0 \\ n_1 &= \epsilon\alpha \quad , \quad n_2 = \epsilon\beta \quad , \quad \text{if} \quad K < 0 \end{aligned}\right\} \tag{4.8.10}$$

The stresses generally vary with time as n_i moves towards these stable steady state values, giving rise to a type of stress relaxation effect. From Eqs. (4.8.3) and (4.8.5), the steady-state stresses are

$$\left.\begin{aligned} t_{11} &= -p + \lambda_1\alpha^2 + (\lambda_2\alpha^2 + 2\lambda_4)\alpha\beta|K| \\ t_{22} &= -p + \lambda_1\beta^2 + (\lambda_2\beta^2 + \lambda_4)\alpha\beta|K| \\ t_{33} &= -p \quad , \quad t_{23} = 0 \quad , \quad t_{31} = 0 \\ t_{12} &= (\lambda_3 + \lambda_2\alpha^2\beta^2 + \lambda_4)K + \begin{cases} \lambda_1\alpha\beta \text{ , if } K > 0 \\ -\lambda_1\alpha\beta \text{ ,if } K < 0 \end{cases} \end{aligned}\right\} \tag{4.8.11}$$

When $\lambda_1 \neq 0$, the shear stress t_{12} does not approach zero as $K \to 0$, which implies that such fluids have an apparent yield stress. Furthermore, the normal stresses t_{11}, t_{22} and t_{33} are generally unequal, so that this theory may, at least qualitatively, explain some of the normal stress effects, such as the Weisenberg effect.

In the steady-state configuration the angle ϕ, determined by $\mathbf{n}$ and the streamlines that are parallel to the x_1 axis, always lies between 0 and $\pi/4$ since

$$0 < \tan \phi = \frac{\beta}{\alpha} = \left(\frac{\mu_3 - 1}{\mu_3 + 1}\right)^{1/2} < 1 \tag{4.8.12}$$

As $\mu_3 \to 1$, the unstable constant solutions given by Eq. (4.8.10) approach the stable solutions given by Eq. (4.8.9), both reducing in the limit to

$$n_1 = \pm 1 \quad , \quad n_2 = n_3 = 0 \tag{4.8.13}$$

Thus **n** can orient itself parallel to the streamlines and remain there. If displaced slightly from this position by a small counterclockwise rotation, **n** will return to its initial position when $K > 0$, its angular velocity going to zero as it approaches its original position. If displaced by a small clockwise rotation, **n** will rotate nearly by an angle π, coming to rest as it becomes parallel to the streamlines. This behavior of **n** can be inferred from Fig. 4.8.1 b.

2. Materials with $0 < \mu_3 < 1$

For $0 < \mu_3 < 1$, Eq. (4.8.4) can be shown to be the same as that derived by Jeffery [Ref. 13] for the motion of an ellipsoid of revolution in a Newtonian fluid, when **n** is taken as the axis of revolution and $\mu_3 = (a^2 - b^2)/(a^2 + b^2)$, $a > b$, where μ_3 is the length of the unequal (major) axis and b the common length of the other two axes. Jeffery's equation has been verified to hold for macroscopic particles. In this case **n** varies periodically with time as well as depending on the solution curve which **n** follows. Details of this solution are discussed by Ericksen [Ref. 1].

4.9 Orientation Induced by Flow

This section explores the phenomenon of orientation induced by flow, an effect that comes about because of the inclusion of microstructure. Only incompressible fluids are considered. The relevant constitutive equations are those given in Eqs. (4.7.7) and (4.7.8). $n^2 = n_r n_r$ will not necessarily be a constant.

Here again the molecular inertia $\rho \ddot{n}_i$ is assumed to be zero, so that Eq. (4.7.8) can be rewritten as

$$\dot{n}_i = f_i(n_k, v_{s,r}) = -w_{ir} n_r + (\mu_1 + \mu_2 d_{rs} n_r n_s) n_i + \mu_3 d_{ir} n_r \tag{4.9.1}$$

where the μ's are arbitrary functions of

$$n^2 = n_r n_r \tag{4.9.2}$$

The stress tensor is given by

$$t_{ij} = -p\delta_{ij} + (\lambda_1 + \lambda_2 d_{rs} n_r n_s) n_i n_j + 2\lambda_3 d_{ij}$$

$$+ 2\lambda_4(d_{ir}n_r n_j + d_{jr}n_r n_i) \tag{4.9.3}$$

where p is an arbitrary pressure and the λ's are functions of n^2. In the absence of body forces, these stresses must satisfy the equations of motion

$$t_{ri,r} = \rho\, \dot{v}_i \tag{4.9.4}$$

For the flows considered in this section, the velocity gradients are constant. Generally, vectors $\mathbf{n}$ satisfying Eq. (4.9.1) will not be constant. For some, but not all of these situations, p can be chosen such that Eq. (4.9.4) is satisfied. When the effects of inertia are negligible, Eq. (4.9.4) can be satisfied by taking solutions of Eq. (4.9.1) which may be time-dependent but do not vary with position. As such, Eq. (4.9.4) will be ignored in this section and the discussion will primarily be concerned with the existence and stability of constant solutions of Eq. (4.9.1), that is, of constant vectors which satisfy

$$f_i(n_k, v_{s,r}) = -w_{ir}n_r + (\mu_1 + \mu_2 d_{rs}n_r n_s)n_i + \mu_3 d_{ir}n_r = 0 \tag{4.9.5}$$

Let $\mathbf{N}$ denote the solution of the steady-state equation given in Eq. (4.9.5), so that $\mathbf{N}$ corresponds to a possible steady-state orientation. To examine its stability let

$$n_i = N_i + m_i \tag{4.9.6}$$

where m_i is a small time-dependent perturbation on $\mathbf{N}$. The objective is then to determine whether such perturbations grow or decay. A substitution from Eq. (4.9.6) in Eq. (4.9.1) gives

$$\dot{N}_i + \dot{m}_i = f_i(N_r + m_r, v_{s,r})$$

$$= f_i(N_k, v_{s,r}) + A_{ir}(N_r, v_{s,r})m_r + \cdots$$

In this equation $\dot{N}_i = 0$ and $f_i(N_r, v_{s,r}) = 0$, since N_i is a steady-state solution. The coefficients A_{ir} in the Taylor expansion of f_i are given by $A_{ir}(N_r, v_{s,r}) = \partial f_i(n_k, v_{q,p})/\partial n_r|_{n_r=N_r}$. If the perturbation m_i is assumed to be small, then the linearized form of the above equations is given by

$$\dot{m}_i = A_{ir}(N_p, v_{q,p})m_r \quad , \quad A_{ir} = \frac{\partial f_i}{\partial n_r}\Big|_{n_r=N_r} \tag{4.9.7}$$

so that from Eq. (4.9.1)

$$A_{ir} = -w_{ir} + (\bar{\mu}_1 + \bar{\mu}_2 d_{pq}N_p N_q)\delta_{ir} + 2\bar{\mu}_2 d_{rs}N_s N_i + \bar{\mu}_3\, d_{ir}$$

$$+ \left[\frac{\partial \bar{\mu}1}{\partial n_r} + \frac{\partial \bar{\mu}_2}{\partial n_r}\, d_{pq}N_p N_q\right] N_i + \frac{\partial \bar{\mu}_3}{\partial n_r}\, d_{is}N_s$$

where $\bar{\mu}_i = \mu_i|_{n_r=N_r}$ and $\partial\bar{\mu}_i/\partial n_r = \partial\mu_i/\partial n_r|_{n_r=N_r}$. Since the μ_i's are functions of n^2, $\partial\mu_i/\partial n_r = (\partial\mu_i/\partial n^2)(\partial n^2/\partial n_r) = 2n_r\partial\mu_i/\partial n^2$, so that

$$A_{ir} = -w_{ir} + (\bar{\mu}_1 + \bar{\mu}_2 d_{pq}N_pN_q)\delta_{ir} + 2\bar{\mu}_2 d_{rs}N_sN_i + \bar{\mu}_3 d_{ir}$$
$$+ 2(\bar{\mu}'_1 + \bar{\mu}'_2 d_{pq}N_pN_q)N_rN_i + 2\bar{\mu}'_3 d_{is}N_sN_r \qquad (4.9.8)$$

where $\bar{\mu}'_i = \partial\mu_i/\partial n^2|_{n_r=N_r}$.

The solutions of the linearized equations, given in Eq. (4.9.7), then determine whether the solution N_i is stable (if m_i decays with time) or unstable (if m_i increases with time). For a complete stability analysis, perturbations of the velocity v_i should also be considered in Eq. (4.9.4). However, this complete problem will not be discussed. As such, some situations which are labelled as being stable may actually turn out to be unstable when the complete problem is considered.

Several special cases will now be considered.

1. The Case $\mathbf{N} = 0$

Assuming that the μ's are well behaved at $n^2 = 0$, Eq. (4.9.5) always has one solution

$$\mathbf{n} = \mathbf{N} = 0 \qquad (4.9.9)$$

For this solution, the stresses given by Eq. (4.9.3) are those for a Newtonian fluid with viscosity $\lambda_3(0)$, and from Eq. (4.9.8) the coefficients in Eq. (4.9.7) are given by

$$A_{ir} = -w_{ir} + \mu_1^0\delta_{ir} + \mu_3^0 d_{ir}\ ,\ \mu_1^0 = \mu_1(0)\ ,\ \mu_3^0 = \mu_3(0) \qquad (4.9.10)$$

Let $B_{ir} = A_{ir} - \frac{1}{3}A_{ss}\delta_{ir}$ be the deviatoric part of $\mathbf{A}$. Then, since $d_{ss} = 0$ for an incompressible fluid,

$$B_{ir} = -w_{ir} + \mu_3^0 d_{ir} \qquad (4.9.11)$$

Also, if M_i is defined by

$$M_i = m_i \exp(-\mu_1^0 t) \qquad (4.9.12)$$

then

$$\dot{M}_i = B_{ir}M_r \qquad (4.9.13)$$

When there is no motion Eq. (4.9.11) shows that $B_{ik} = 0$, so that M_i is constant and $m_i = C_i\exp(\mu_1^0 t)$, where the C_i's are constants. The fluid under consideration will be assumed to be unoriented at rest, in the sense that $\mathbf{N} = 0$ is a stable solution, which is equivalent to requiring

$$\mu_1^0 < 0 \tag{4.9.14}$$

To determine the conditions under which $\mathbf{N} = 0$ will be stable when there is flow, the conditions under which all solutions of $\dot{m}_i = A_{ir} m_r$ approach zero must be investigated. Since $\dot{m}_i = A_{ir} m_r$ is a system of linear homogeneous equations with constant coefficients, its solutions are expressible as linear combinations of complex exponential solutions with coefficients that are generally constant, but sometimes polynomials in t, so that for stability, all solutions of Eq. (4.9.13) of the form

$$M_i = C_i \exp(\alpha t) \tag{4.9.15}$$

where C_i and α are complex constants, must be such that

$$\mu_1^0 + Real\ \alpha < 0 \tag{4.9.16}$$

Now for Eq. (4.9.15) to be a solution of Eq. (4.9.13) $B_{ir} M_r = \dot{M}_i = \alpha C_i \exp(\alpha t) = \alpha M_i$, which results in the characteristic equation $(B_{ir} + \alpha \delta_{ir}) C_r = 0$. The admissible values of α are thus the roots of

$$y \equiv \det(\mathbf{B} - \alpha \mathbf{I}) = -\alpha^3 - II\alpha + III = 0 \tag{4.9.17}$$

where

$$II = -\frac{1}{2} B_{ir} B_{ri} \quad , \quad III = \det\mathbf{B} \tag{4.9.18}$$

are the invariants of $\mathbf{B}$. Since the coefficient of α^2 in Eq. (4.9.17) is zero, the three roots α_1, α_2 and α_3 must satisfy

$$\alpha_1 + \alpha_2 + \alpha_3 = 0 \tag{4.9.19}$$

Consider first the case when all the three roots are real, so that

$$27 III^2 + 4 II^3 \leqslant 0 \tag{4.9.20}$$

If the roots are arranged such that $\alpha_1 \leqslant \alpha_2 \leqslant \alpha_3$, then Eq. (4.9.16) reduces to

$$\mu_1^0 + \alpha_3 < 0 \tag{4.9.21}$$

This criterion is awkward to use when the cubic equation is not easily solved. Now an examination of the graph of $y = y(\alpha)$, given by

Eq. (4.9.17), will show that it has a maximum at $\alpha = \beta = (-II/3)^{1/2}$. Also, any value of α greater than this, for which $y(\alpha) < 0$, must exceed the largest root α_3. An application of this reasoning to the value $\alpha = -\mu_1^0$ will show that Eq. (4.9.21) is equivalent to the two conditions

$$\left.\begin{aligned} y(-\mu_1^0) &= (\mu_1^0)^3 + II\mu_1^0 + III < 0 \\ -\mu_1^0 &> \beta \end{aligned}\right\} \tag{4.9.22}$$

as long as Eq. (4.9.20) holds.

When Eq. (4.9.20) is violated, Eq. (4.9.17) will have one real and two conjugate complex roots. Because of Eq. (4.9.19), these roots must have the form $a + ib$, $a - ib$ and $-2a$, where a and b are real. Equation (4.9.16) then reduces to

$$\mu_1 + a < 0 \quad \text{if} \quad a \geqslant 0 \tag{4.9.23}$$

$$\mu_1 - 2a < 0 \quad \text{if} \quad a < 0 \tag{4.9.24}$$

Now from Eq. (4.9.17)

$$III = \alpha_1\alpha_2\alpha_3 = -2a(a^2 + b^2) \tag{4.9.25}$$

Since III can be calculated easily from $III = \det\mathbf{B}$, Eq. (4.9.25) can be used for determining the sign of a without finding the roots of Eq. (4.9.17). When $III \leqslant 0$, $a \geqslant 0$ and the inequality in Eq. (4.9.23) is satisfied provided that

$$y(2\mu_1^0) = -8(\mu_1^0)^3 - 2II\mu_1^0 + III > 0 \tag{4.9.26}$$

Similarly, when $III > 0$, Eq. (4.9.24) can be shown to be equivalent to

$$y(-\mu_1^0) = (\mu_1^0)^3 + II\mu_1^0 + III < 0 \tag{4.9.27}$$

Thus, to determine whether $N = 0$ is a stable solution for a given motion of a given material, is a straightforward matter. Equations (4.9.22), (4.9.26) and (4.9.27) can be used to infer that $\mathbf{N} = 0$ will be stable if the velocity gradients are sufficiently small. Also, solutions can be constructed for which $\mathbf{N} = 0$ is unstable. Ericksen [Ref. 6] has shown that this generally occurs in simple shear at sufficiently high rates of shear.

2. *Irrotational Motion*

Since $\mathbf{D}$ is symmetric, a rectangular Cartesian coordinate system in which it has the diagonal form

$$\mathbf{D} = \begin{bmatrix} d_1 & 0 & 0 \\ 0 & d_2 & 0 \\ 0 & 0 & d_3 \end{bmatrix}$$

can always be chosen with $d_1 + d_2 + d_3 = 0$, since only incompressible fluids are being considered. For irrotational motions $\mathbf{W} = 0$, so that in this coordinate system the roots of Eq. (4.9.17) are $\mu_3^0 d_1$, $\mu_3^0 d_2$ and $\mu_3^0 d_3$. These may be arranged in the order

$$\mu_3^0 d_1 \leqslant \mu_3^0 d_2 \leqslant \mu_3^0 d_3 \tag{4.9.28}$$

Equation (4.9.21) then shows that $\mathbf{N} = 0$ is stable provided

$$\mu_1^0 + \mu_3^0 d_3 < 0 \tag{4.9.29}$$

In this case the fluid behaves like a Newtonian fluid. When Eq. (4.9.29) is violated, so that $\mathbf{N} = 0$ is unstable, this solution is of little interest. An alternative would be to look for a steady-state stable solution with $\mathbf{N} \neq 0$, when Eq. (4.9.29) fails.

To investigate the possibility of a steady-state stable solution with $\mathbf{N} \neq 0$, $\mathbf{N}$ must satisfy $f_i(N_r, v_{s,r}) = 0$, that is

$$(\mu_1 + \mu_2 d_{rs} N_r N_s) N_i + \mu_3 d_{ir} N_r = 0 \tag{4.9.30}$$

Assuming that $\mu_3 \neq 0$, this implies that $\mathbf{N}$ must be a proper vector of $\mathbf{D}$ corresponding to some proper value which is, say d_3, so that

$$d_{ir} N_r = d_3 N_i \quad , \quad d_{rs} N_r N_s = d_3 N^2 \quad , \quad \mathbf{N} = (0,0,N)$$

To avoid a possible loss of generality, the ordering in Eq. (4.9.28) will not be assumed for the present. Equation (4.9.30) then reduces to

$$\bar{\mu}_1 + \bar{\mu}_3 d_3 + \bar{\mu}_2 d_3 N^2 = 0 \tag{4.9.31}$$

which can be used for determining N^2 in terms of d_3. Assuming that an $\mathbf{N}$ satisfying all these conditions exists then, from Eq. (4.9.8), A_{ir} in $\dot{m}_i = A_{ir} m_r$ is given by

$$A_{ir} = (\bar{\mu}_1 + \bar{\mu}_2 d_3 N^2)\delta_{ir} + \bar{\mu}_3 d_{ir}$$
$$+ 2(\bar{\mu}_2 d_3 + \bar{\mu}'_1 + \bar{\mu}'_2 d_3 N^2 + \bar{\mu}'_3 d_3) N_i N_r$$

Which, upon using Eq. (4.9.31), can be rewritten as

$$A_{ir} = -\bar{\mu}_3 d_3 \delta_{ir} + \bar{\mu}_3 d_{ir} + 2(\bar{\mu}_2 d_3 + \bar{\mu}'_1 + \bar{\mu}'_2 d_3 N^2 + \bar{\mu}'_3 d_3) N_i N_r \tag{4.9.32}$$

The stability equation, $\dot{m}_i = A_{ir} m_r$, then reduces to

$$\left.\begin{aligned} \dot{m}_1 &= \bar{\mu}_3(d_1 - d_3)\, m_1 \\ \dot{m}_2 &= \bar{\mu}_3(d_2 - d_3)\, m_2 \\ \dot{m}_3 &= 2N^2\chi(N^2, d_3)\, m_3 \end{aligned}\right\} \tag{4.9.33}$$

where $\bar{\mu}_3 = \mu_3(N^3)$ and

$$\begin{aligned} \chi(N^2, d_3) &= (\bar{\mu}'_1 + \bar{\mu}'_2 d_3 N^2 + \bar{\mu}'_3 d_3 + \bar{\mu}_2 d_3) \\ &= \frac{\partial}{\partial n^2}\, (\mu_1 + \mu_2 n^2 d_3 + \mu_3 d_3)|_{n^2 = N^2} \end{aligned} \tag{4.9.34}$$

The stability criteria are then given by

$$\bar{\mu}_3(d_1 - d_3) < 0 \tag{4.9.35}$$

$$\bar{\mu}_3(d_2 - d_3) < 0 \tag{4.9.36}$$

$$N^2\chi(N^2, d_3) < 0 \tag{4.9.37}$$

If μ_3 is assumed to always be of one sign, then Eqs. (4.9.33) and (4.9.34) determine which proper value of $\mathbf{D}$ is to be d_3, the ordering given by Eq. (4.9.28) being satisfactory.

From Eq. (4.9.29), a critical value d of d_3 is given by

$$d = -\mu_1^0/\mu_3^0$$

Suppose, as is suggested by Eq. (4.9.37), that

$$\chi(0, d) < 0 \tag{4.9.38}$$

and note that Eq. (4.9.31) is satisfied when $d_3 = d$, $N^2 = 0$. For d_3 sufficiently close to d, Eq. (4.9.38) together with the implicit function theorem guarantees that there will be a unique solution of Eq. (4.9.31)

$$N^2 = f(d_3) \quad , \quad f(d) = 0 \tag{4.9.39}$$

Also, Eq. (4.9.31) can be used to show that

$$\frac{1}{\mu_3^0}\, \frac{df}{dd_3}|_{d_3 = d} = -\, \frac{1}{\chi(0, d)} > 0 \tag{4.9.40}$$

Then a use of this equation shows that below the critical value, where Eq. (4.9.29) holds, $N^2 < 0$, so that $\mathbf{N}$ is not a real vector. Slightly above the critical value, $N^2 > 0$, resulting in a real vector. Except when d_3 is one of a pair of equal proper values, the direction of $\mathbf{N}$ is uniquely determined as that of the corresponding proper vector. Equation (4.9.38) shows that this solution will satisfy Eq. (4.9.37) for d_3 sufficiently close but not equal to d. When all these assumptions are met, there will be a steady state $\mathbf{N} = \mathbf{N}(d_3)$ such that

$$\mathbf{N} = 0 \quad \text{if} \quad \mu_1^0 + \mu_3^0 d_3 \leqslant 0$$

$$\mathbf{N} \neq 0 \quad \text{if} \quad \mu_1^0 + \mu_3^0 d_3 > 0$$

When $\mathbf{N} \neq 0$, the stresses will generally differ qualitatively from those of a Newtonian fluid, as can be seen from Eq. (4.9.3).

As an example let $\mu_3 > 0$ and consider the simple extension $d_3 = -2d_1 = -2d_2 > 0$. In this case, the unoriented condition is stable provided the rate of extension d_3 is small enough such that $\mu_1^0 + \mu_3^0 d_3 < 0$. At higher rates, orientation will occur with $\mathbf{n}$ tending to be parallel to the direction of d_3. For the same material in simple compression, $d_1 = -2d_2 = -2d_3 < 0$, so that the stability condition, given by Eq. (3.9.46), fails. At relatively high rates of extension, $\mathbf{n}$ tends to be perpendicular to the axis of compression.

The paper of Ericksen [Ref. 6], from which the results of this section have been abstracted, goes on to consider two additional types of flows: plane flows for which

$$\mathbf{D} = \begin{bmatrix} 0 & d & 0 \\ d & 0 & 0 \\ 0 & 0 & 0 \end{bmatrix}, \quad \mathbf{W} = \begin{bmatrix} 0 & w & 0 \\ -w & 0 & 0 \\ 0 & 0 & 0 \end{bmatrix}$$

and simple shear, which is a special case of plane flows with $d = w = \frac{1}{2}\dot{\gamma}$, where $\dot{\gamma}$ is the rate of shear.

For plane flows Ericksen has shown that $\mathbf{N} = 0$ is stable provided that $(\mu_1^0)^2 > (\mu_3^0 d)^2 - w^2$, so that the vorticity has a stabilizing effect. Here again, if this condition is violated, a stable $\mathbf{N} \neq 0$ is possible under certain conditions.

4.10 Poiseuille Flow Through Circular Pipes

In this section steady Poiseuille flow of an anisotropic fluid through a circular pipe is considered. It is assumed that the fluid is incompressible and that $\mathbf{n}$ is a unit vector. The "molecular inertia" $\rho \ddot{n}_i$ is neglected. The relevant equations are then given by Eqs. (4.7.10), (4.7.11) and Cauchy's first law of motion $\rho a_i = t_{ji,j}$.

Introducing cylindrical polar coordinates (r,θ,z), such that the z axis is the axis of the pipe, the velocity field may be assumed to have the form

$$u = 0 \quad , \quad v = 0 \quad , \quad w = w(r) \tag{4.10.1}$$

where u, v and w are the components of the velocity along the r, θ and z directions. For this case, the only nonzero components of **D** and **W** are given by

$$d_{zr} = \frac{1}{2}w' \quad , \quad w_{zr} = -\frac{1}{2}w' \tag{4.10.2}$$

where a prime denotes differentiation with respect to r. From Eqs. (4.7.10) the stresses are then given by

$$\left.\begin{aligned}
t_{rr} &= -p + Xn_r^2 + 2\lambda_4 w' n_r n_z \\
t_{\theta\theta} &= -p + Xn_\theta^2 \\
t_{zz} &= -p + Xn_z^2 + 2\lambda_4 n_r n_z \\
t_{\theta z} &= Xn_\theta n_z + \lambda_4 w' n_r n_\theta \\
t_{zr} &= Xn_z n_r + [\lambda_4(1 - n_\theta^2) + \lambda_3] w' \\
t_{r\theta} &= Xn_r n_\theta + \lambda_4 w' n_\theta n_z
\end{aligned}\right\} \tag{4.10.3}$$

where $X = \lambda_1 + \lambda_2 w' n_r n_z$. Similarly, Eq. (4.7.11) reduces to

$$\left.\begin{aligned}
\dot{n}_r &= \frac{\partial n_r}{\partial t} + w\frac{\partial n_r}{\partial z} = \frac{1}{2}w' n_z(\mu_3 - 1 - 2\mu_3 n_r^2) \\
\dot{n}_\theta &= \frac{\partial n_\theta}{\partial t} + w\frac{\partial n_\theta}{\partial z} = -\mu_3 w' n_r n_\theta n_z \\
\dot{n}_1 &= \frac{\partial n_z}{\partial t} + w\frac{\partial n_z}{\partial z} = -\frac{1}{2}w' n_r(\mu_3 + 1 - 2\mu_3 n_z^2)
\end{aligned}\right\} \tag{4.10.4}$$

Since the stresses are only functions of r and z, Cauchy's first law, $\rho a_i = t_{ri,r}$, reduces to

$$\left.\begin{aligned}
&\frac{\partial t_{rr}}{\partial r} + \frac{\partial t_{zr}}{\partial z} + \frac{1}{r}(t_{rr} - t_{\theta\theta}) = 0 \\
&\frac{\partial t_{r\theta}}{\partial r} + \frac{\partial t_{z\theta}}{\partial z} + \frac{2}{r}t_{r\theta} = 0 \\
&\frac{\partial t_{rz}}{\partial r} + \frac{\partial t_{zz}}{\partial z} + \frac{1}{r}t_{rz} = 0
\end{aligned}\right\} \tag{4.10.5}$$

The fluid behaves as a Bingham material provided

$$|\mu_3| > 1 \quad , \quad \lambda_1 \neq 0 \tag{4.10.6}$$

This equation will therefore be assumed to hold. Equation (4.10.4) can then be shown to have the steady state solution

$$n_r = \sin\phi \quad , \quad n_\theta = 0 \quad , \quad n_z = \cos\phi \tag{4.10.7}$$

where ϕ is a constant angle given by

$$\tan^2\phi = (\mu_3 - 1)/(\mu_3 + 1) \tag{4.10.8}$$

which lies in the range

$$\left.\begin{array}{ll} 0 < \phi < \pi/r & \textit{when } w' > 0,\ \mu_3 > 1 \\ 3\pi/4 < \phi < \pi & \textit{when } w' < 0,\ \mu_3 > 1 \\ \pi/4 < \phi < \pi/2 & \textit{when } w' < 0,\ \mu_3 < -1 \\ \pi/2 < \phi < 3\pi/4 & \textit{when } w' > 0,\ \mu_3 < -1 \end{array}\right\} \tag{4.10.9}$$

The theory treats $\mathbf{n}$ and $-\mathbf{n}$ as indistinguishable, which means that there is no loss of generality in taking $0 \leqslant \phi \leqslant \pi$, as is done here.

1. Steady-State Flow

The stresses for the steady-state are then

$$\left.\begin{array}{l} t_{rr} = -p + \lambda_1 \sin^2\phi + Aw' \\ t_{\theta\theta} = -p \\ t_{zz} = -p + \lambda_1 \cos^2\phi + Bw' \\ t_{\theta z} = 0\ ,\ t_{zr} = C + Dw'\ ,\ t_{r\theta} = 0 \end{array}\right\} \tag{4.10.10}$$

where the constants A, B, C and D are given by

$$\left.\begin{array}{l} 2A = (\lambda_2 \sin^2\phi + 2\lambda_4)\sin 2\phi \\ 2B = (\lambda_2 \cos^2\phi + 2\lambda_4)\sin 2\phi \\ 2C = \lambda_1 \sin 2\phi \\ 4D = 4(\lambda_3 + \lambda_4) + \lambda_2 \sin^2 2\phi \end{array}\right\} \tag{4.10.11}$$

Note that $t_{rz} \rightarrow C$ as $w' \rightarrow 0$, so that C can be interpreted as an apparent yield shear stress. Similarly, D is an apparent fluidity or "viscosity." The constants A, B and C reverse their signs with w', so that places where $w' = 0$ must be treated with care.

Integration of Eqs. (4.10.5) a, c, with the help of Eqs. (4.10.10), then gives

$$p = \lambda_1 \sin^2\phi + Aw' + \int \frac{1}{r}(\lambda_1 \sin^2\phi + Aw')\,dr - Ez \tag{4.10.12}$$

$$t_{rz} = C + Dw' = -\frac{1}{2}Er + \frac{F}{r} \tag{4.10.13}$$

where E and F are constants of integration.

Since the stresses are expected to do non-negative work in deforming the fluid, $t_{rz}w' = (C + Dw')w' \geqslant 0$. For this inequality to hold, the use of Eqs. (4.10.11) c, d, Eqs. (4.10.8) and (4.10.9) shows that the following conditions must be satisfied

$$\left.\begin{array}{l} \lambda_1 \geqslant 0\ ,\ \lambda_2(\mu_3^2 - 1) + 4\mu_3^2(\lambda_3 + \lambda_4) \geqslant 0 \ \textit{when}\ \mu_3 > 1 \\ \lambda_1 \leqslant 0\ ,\ \lambda_2(\mu_3^2 - 1) + 4\mu_3^2(\lambda_3 + \lambda_4) \geqslant 0 \ \textit{when}\ \mu_3 < -1 \end{array}\right\} \tag{4.10.14}$$

The assumptions that have been made rule out the possibility of singularities at $r = 0$. That plug flow is possible, just as in Bingham materials, will now be shown.

2. Plug Flow

Now consider the possibility of a solution $w = w_0 = \textit{constant}$, so that $w' = 0$, which does not have a singularity at $r = 0$. Then from Eq. (4.10.4) $\dot{n}_r = \dot{n}_\theta = \dot{n}_z = 0$. For steady flow this implies that $\partial n_i/\partial z = 0$, and, since the flow is assumed to be axisymmetric,

$$n_r = n_r(r) \quad , \quad n_\theta = n_\theta(r) \quad , \quad n_z = n_z(r) \tag{4.10.15}$$

so that all the terms in the stresses given by Eq. (4.10.3), with the exception of p which is a function of r and z, will be functions of r only. Then Eqs. (4.10.5) b and (4.10.3) f give $r^2 t_{r\theta} = \lambda_1 n_r n_\theta = G$, *a* constant. For $t_{r\theta}$ to be finite as $r \to 0$, $G = 0$, so that

$$n_r n_\theta = 0 \tag{4.10.16}$$

Equation (4.10.3) e shows that $n_r = 0$ implies $t_{rz} = 0$, which is not true, in general, for pipe flow. Since $\mathbf{n}$ is a unit vector, a possible solution, which satisfies Eq. (4.10.16), is then

$$n_r = \sin\psi \quad , \quad n_\theta = 0 \quad , \quad n_z = \cos\psi \tag{4.10.17}$$

From Eq. (4.10.3) that the stresses are then given by

$$\left.\begin{aligned} t_{rr} &= -p + \lambda_1 \sin^2\psi \\ t_{\theta\theta} &= -p \ , \ t_{zz} = -p + \lambda_1 \cos^2\psi \\ t_{\theta z} &= 0 \ , \ t_{zr} = \frac{1}{2} \lambda_1 \sin 2\psi \ , \ t_{r\theta} = 0 \end{aligned}\right\} \tag{4.10.18}$$

Integration of Eqs. (4.10.5) a, c, subject to the condition that t_{zr} remain finite at $r \to 0$, gives

$$p = \lambda_1 \sin^2\psi + \lambda_1 \int \frac{1}{r} \sin^2\psi \, dr - Hz \tag{4.10.19}$$

$$2t_{rz} = \lambda_1 \sin 2\psi = -Hr \tag{4.10.20}$$

where H is a constant of integration. Because of the indistinguishability of $\mathbf{n}$ and $-\mathbf{n}$, there is no loss of generality in taking $0 \leqslant \psi < \pi$, so that from Eq. (4.10.20)

$$\sin^2\psi = \frac{1}{2}\left\{1 - \left[1 - (Hr/\lambda_1)^2\right]^{1/2}\right\} \tag{4.10.21}$$

with

$$\left.\begin{aligned} 0 \leqslant \psi \leqslant \pi/4 \quad &\text{if} \quad H/\lambda_1 < 0 \\ 3\pi/4 \leqslant \psi \leqslant \pi \quad &\text{if} \quad H/\lambda_1 > 0 \end{aligned}\right\} \tag{4.10.22}$$

Equations (4.10.17), (4.10.21) and (4.10.22) then give a solution of the type w = constant, which is valid for

$$0 \leqslant r \leqslant R_1 = |\lambda_1/H| \tag{4.10.23}$$

3. Combined Flow

Two sets of solutions have now been obtained: a constant solution $w = w_0$ valid in a region $0 \leqslant r \leqslant R_1$, and a solution with variable w which was obtained earlier. The next step is to investigate the possibility of a combined solution consisting of a variable velocity $w(r)$ for $r \geqslant R_2$, and a plug flow solution $w \equiv w_0$ for $r \leqslant R_2$. The velocity and the stress vector must, of course, be continuous at $r = R_2$. That is, the solutions must be matched so as to make t_{rr}, $t_{r\theta}$, t_{rz} and w continuous at $r = R_2$. For the moment R_2 represents any value in the range $0 \leqslant r \leqslant R_1$. Equations (4.10.10) f and (4.10.18) show that the continuity condition on $t_{r\theta}$ is automatically satisfied since $t_{r\theta}$ is identically zero for both the solutions.

For plug flow, use of Eqs. (4.10.18) a, (4.10.19) and (4.10.21) gives

$$t_{rr} = -\frac{1}{2}\int_0^r \frac{1}{x}\left\{1 - \left[1 - (Hx/\lambda_1)^2\right]^{1/2}\right\} dx + Hz + K \ , \ r \leqslant R_2 \quad (4.10.24)$$

where K is a constant of integration. In the variable velocity region, Eqs. (4.10.10) a and (4.10.12) show that

$$t_{rr} = -\int_{R_2}^r \frac{1}{x}(\lambda_1 \sin^2\phi + Aw')\,dx + Ez + L \ , \ r \geqslant R_2 \quad (4.10.25)$$

For Eqs. (4.10.24) and (4.10.25) to agree at $r = R_2$ for arbitrary z,

$$E = H \ , \ L = -\frac{1}{2}\lambda_1\int_0^{R_2} \frac{1}{x}\left\{1 - \left[1 - (Hx/\lambda_1)^2\right]^{1/2}\right\} dx + K \quad (4.10.26)$$

Use of Eqs. (4.10.26) a, (4.10.13) and (4.10.20) shows that, for t_{rz} to be continuous at $r = R_2$, $F = 0$, so that

$$t_{rz} = -\frac{1}{2}Hr \quad (4.10.27)$$

Then, from Eqs. (4.10.13) and (4.10.27)

$$w = -\frac{1}{4D}(Hr^2 + 4Cr) + M \quad (4.10.28)$$

where M is an arbitrary constant. For w to be continuous throughout

$$w_0 = -\frac{1}{4D}(HR_2^2 + 4CR_2) + M \quad (4.10.29)$$

As long as the specified conditions are satisfied, a satisfactory combination of the solutions, which is valid for all r, has been obtained. Thus there would seem to be an abundance of solutions to the Poiseuille flow problem. On the basis of a plausibility argument, most of the solutions will now be ruled out on the grounds that they describe unstable solutions.

For the plug flow solution, the limiting value $\bar{\mathbf{n}}$ of $\mathbf{n}$, for $r \rightarrow R_2$, $r < R_2$, as a function of R_2, is given by

$$\bar{n}_r = \sin\psi(R_2) \quad , \quad \bar{n}_\theta = 0 \quad , \quad \bar{n}_z = \cos\psi(R_2) \quad (4.10.30)$$

The limiting value of the velocity gradient, as $r \rightarrow R_2$ from above, is given from Eq. (4.10.28) by

$$w'(R_2) = -\frac{1}{2D}(HR_2 + 2C) \quad (4.10.31)$$

which will, in general, tend to rearrange the structure represented by $\mathbf{n}$. That is, if $\bar{\mathbf{n}}$ is regarded as an initial value of $\mathbf{n}$ in Eq. (4.10.4), then $w'(R_2)$ would cause $\mathbf{n}$ to change in accordance with

$$\dot{n}_r = \frac{1}{2} w'(R_2)\, n_z(\mu_3 - 1 - 2\mu_3 n_r^2)$$

$$\dot{n}_z = \frac{1}{2} w'(R_2)\, n_r(\mu_3 - 1 - 2\mu_3 n_z^2)$$

n_θ remaining zero, which is expected to make the interface unstable. Exceptions might occur if

$$\psi(R_2) = \phi \tag{4.10.32}$$

when $\mathbf{n}$ coincides with the stable, steady state value of $\mathbf{n}$ in the flow exterior to $r = R_2$, or if

$$w'(R_2) = 0 \tag{4.10.33}$$

which implies that $\dot{\mathbf{n}} = 0$. If Eq. (4.10.32) holds, then Eqs. (4.10.20) and (4.10.11) c give

$$R_2 = -\frac{\lambda_1}{H} \sin 2\psi(R_2) = -\frac{\lambda_1}{H} \sin 2\phi = -\frac{2C}{H} \tag{4.10.34}$$

so that it follows from Eq. (4.10.31) that Eq. (4.10.33) holds. Hence Eq. (4.10.33) must hold and, at the interface

$$t_{rz} = -\frac{1}{2} HR_2 = C$$

Thus the shear stress takes on its apparent yield value at the interface, which is in accord with the analyses of Poiseuille flow of Bingham materials, where, at the boundary of the rigidly moving plug, the velocity gradient vanishes and the shear stress takes on its yield value.

If Eq. (4.10.33) holds, then Eq. (4.10.34) must hold, so that either Eq. (4.10.32) holds or

$$\psi(R_2) = \frac{\pi}{2} - \phi \tag{4.10.35}$$

Then Eqs. (4.10.22) and (4.10.9) show that Eq. (4.10.32) but not Eq. (4.10.34) can hold if $\mu_3 > 1$, while Eq. (4.10.34) but not Eq. (4.10.32) can hold if $\mu_3 < -1$.

The argument used here does not settle the question of the stability of these flows. For example, it says nothing about the stability of the noted exceptional cases.

4. Poiseuille Flow

The solutions that have been developed so far will now be used for describing the steady flow through an infinitely long pipe of radius R due to a given pressure gradient $-\partial p/\partial z$. Now Eqs. (4.10.10) c and (4.10.18) c imply that $-\partial p/\partial z = \partial t_{zz}/\partial z$. A use of Eqs. (4.10.5) c and (4.10.27) then results in

$$H = -\frac{\partial p}{\partial z} \tag{4.10.36}$$

As usual, the fluid will be assumed to adhere to the wall at $r = R$.

The consequences of the plausibility argument used above will be assumed. Also, the pressure gradient H will be assumed to be sufficiently large to make

$$R_2 = -2CH < R \tag{4.10.37}$$

Using the no-slip condition to evaluate the constant M in Eqs. (4.10.28) and (4.10.29)

$$w = -\frac{1}{4D}(r - R)[H(r + R) + 4C] \quad , \quad r \geqslant R_2$$

$$w = w_0 = -\frac{1}{4D}(R_2 - R)[H(R_2 + R) + 4C] \quad , \quad r \leqslant R_2$$

where C and D are given by Eqs. (4.10.11) c and d.

Now Eq. (4.10.34) requires that $-\lambda_1 H \sin 2\phi > 0$. An appropriate value of ϕ can then be determined by using Eqs. (4.10.14) a or b and Eqs. (4.10.8) and (4.10.9). For a given material, the velocity distribution is thus uniquely determined as long as Eq. (4.10.37) holds. With the understanding that C and D are interpretable as the yield stress and fluidity, respectively, this velocity distribution is indistinguishable from that for a Bingham material.

If the pressure gradient, $H = -\partial p/\partial z$, is sufficiently low, Eq. (4.10.37) will fail, as $R_2 > R$. A reasonable interpretation is that the material remains at rest. In other words, the plug flow solution $w_0 = 0$ applies. Then, from Eqs. (4.10.18) to (4.10.22),

$$\left.\begin{aligned}
&t_{rr} = Y \ , \ t_{\theta\theta} = Y - \frac{1}{2}\lambda_1\left\{1 - \left[1 - (Hr/\lambda_1)^2\right]^{1/2}\right\} \\
&t_{zz} = Y + \lambda_1\left[1 - (Hr/\lambda_1)^2\right]^{1/2} \\
&t_{\theta z} = 0 \ , \ t_{zr} = -\frac{1}{2}Hr \ , \ t_{r\theta} = 0
\end{aligned}\right\} \tag{4.10.38}$$

where $Y = -\frac{1}{2}\lambda_1 \int_0^r \frac{1}{x}\left\{1 - \left[1 - (Hx/\lambda_1)^2\right]^{1/2}\right\} dx + Hz + K$. When flow occurs, these stresses hold for $0 \leqslant r \leqslant R_2$. Similarly, for $R_2 \leqslant r \leqslant R$

$$\left.\begin{aligned} &t_{rr} = Z\ ,\ t_{\theta\theta} = Z - \lambda_1 \sin^2\phi + \frac{A}{2D}(Hr + 2C) \\ &t_{zz} = Z + \lambda_1 \cos 2\phi + \frac{A - B}{2D}(Hr + 2C) \\ &t_{\theta z} = 0\ ,\ t_{zr} = -\frac{1}{2}Hr\ ,\ t_{r\theta} = 0 \end{aligned}\right\} \qquad (4.10.39)$$

where A, B, C and D are the constants given in Eq. (4.10.11),

$$Z = -\int_{R_2}^r \frac{1}{x}(\lambda_1 \sin^2\phi + Aw')\, dx + Ez + L$$

and L is given by Eq. (4.10.26) b. On integration, this results in

$$Z = \left[\frac{AC}{D} - \lambda_1 \sin^2\phi\right] \ln(r/R_2) + \frac{AH}{2D}(r - R_2) + Ez + L$$

Except for K, which remains arbitrary, all constants are uniquely determined as functions of $H = -\partial p/\partial z$. Changing K changes the stress by a uniform hydrostatic pressure, which has no effect on incompressible materials.

The stresses given by Eqs. (4.10.38) or (4.10.39) differ from those predicted by Oldroyd's [Ref. 14] equation for Bingham materials, which imply that $t_{rr} = t_{\theta\theta} = t_{zz}$. However, in terms of quantities generally discussed in connection with Poiseuille flow of Bingham materials, the two theories yield indistinguishable predictions.

4.11 Cylindrical Couette Flow

In this section, the steady incompressible flow of an anisotropic fluid between two coaxial right cylinders, of radii R_1 and $R_2 > R_1$, which are rotating about their common axis with constant angular velocities Ω_1 and Ω_2, will be considered. The main assumptions are the same as those made in the beginning of Section (4.10). Thus, molecular inertia $\rho \ddot{n}_i$ will be assumed to be zero, and **n** to be a unit vector. The relevant equations are then given by Eqs. (4.7.10), (4.7.11) and Cauchy's first law $\rho a_i = t_{ri,r}$.

For the problem under consideration, the appropriate velocity field is of the form

$$u = 0 \quad , \quad v = v(r) \quad , \quad w = 0 \tag{4.11.1}$$

and the boundary conditions are

$$v(R_1) = R_1\Omega_1 \quad , \quad v(R_2) = R_2\Omega_2 \tag{4.11.2}$$

The only nonzero components of **D** and **W** are then given by

$$d_{r\theta} = \frac{1}{2}(v' - v/r) = \frac{1}{2}\gamma \quad , \quad w_{r\theta} = \frac{1}{2}(v' + v/r)$$

where a prime denotes differentiation with respect to r. $\gamma = \gamma(r)$ is then the rate of shear strain. Using Eq. (4.7.10), the stresses are then given by

$$\left.\begin{aligned}
t_{rr} &= -p + \lambda_1 n_r^2 + \gamma(\lambda_2 n_r^2 + 2\lambda_4) n_r n_\theta \\
t_{\theta\theta} &= -p + \lambda_1 n_\theta^2 + \gamma(\lambda_2 n_\theta^2 + 2\lambda_4) n_r n_\theta \\
t_{zz} &= -p + \lambda_1 n_z^2 + \gamma(\lambda_2 n_z^2) n_r n_\theta \\
t_{\theta z} &= \lambda_1 n_\theta n_z + \gamma(\lambda_2 n_\theta^2 + \lambda_4) n_r n_z \\
t_{zr} &= \lambda_1 n_z n_r + \gamma(\lambda_2 n_r^2 + \lambda_4) n_\theta n_z \\
t_{r\theta} &= \lambda_1 n_r n_\theta + \gamma[\lambda_3 + \lambda_2 n_r^2 n_\theta^2 + \lambda_4(n_r^2 + n_\theta^2)]
\end{aligned}\right\} \tag{4.11.3}$$

where $\gamma(r) = (v' - v/r)$.

On assuming that $\mathbf{n} = \mathbf{n}(r,t)$, Eqs. (4.7.11) reduce to

$$\left.\begin{aligned}
\frac{\partial n_r}{\partial t} &= -\frac{1}{2}\gamma n_\theta + \frac{1}{2}\gamma(\mu_3 - 2\mu_3 n_r^2) n_\theta \\
\frac{\partial n_\theta}{\partial t} &= \frac{1}{2}\gamma n_r + \frac{1}{2}\gamma(\mu_3 - 2\mu_3 n_\theta^2) n_r \\
\frac{\partial n_z}{\partial t} &= -\gamma\mu_3 n_r n_\theta n_z
\end{aligned}\right\} \tag{4.11.4}$$

where $n_r^2 + n_\theta^2 + n_z^2 = 1$. Also, since the stresses are only functions of r and t, the equations of motion, $\rho a_i = t_{ri,r}$, reduce to

$$\left.\begin{aligned}
&\frac{\partial t_{rr}}{\partial r} + \frac{1}{r}(t_{rr} - t_{\theta\theta}) = -\frac{\rho v^2}{r} \\
&\frac{\partial t_{r\theta}}{\partial r} + \frac{2}{r} t_{r\theta} = 0 \\
&\frac{\partial t_{rz}}{\partial r} + \frac{1}{r} t_{rz} = 0
\end{aligned}\right\} \tag{4.11.5}$$

On integration, Eqs. (4.11.4) can be shown to have the solution

$$\left.\begin{aligned} n_r^2 &= Kn_\theta^2 + L \\ n_\theta^2 &= \frac{(\mu_3+1)^{1/2}\cosh[(\mu+1)^{1/2}(\frac{1}{2}\gamma t+\delta)] - L}{K + 2\mu_3[L/(\mu_3+1)]^{1/2}\cosh[(\mu_3+1)^{1/2}(\frac{1}{2}\gamma t+\delta)]} \\ n_z^2 &= \beta(\mu_3 - 2\mu_3 n_r^2 - 1) \end{aligned}\right\} \tag{4.11.6}$$

where δ, K, and β are arbitrary functions of r only, and L is given by

$$L = \frac{1}{\mu_3}[\frac{1}{2}(1-K)(1+\mu_3) - 1]$$

A further assumption that $|\mu_3| > 1$ then results in the above solution approaching the steady-state solution

$$n_r = \sin\phi, \quad n_\theta = \cos\phi, \quad n_1 = 0 \tag{4.11.7}$$

as $t \to \infty$, where ϕ is the constant angle determined by

$$\tan^2\phi = (\mu_3 - 1)/(\mu_3 + 1) \tag{4.11.8}$$

and lies in the range

$$\left.\begin{array}{ll} 0 < \phi < \pi/4 & \textit{when } \gamma > 0,\ \mu_3 > 1 \\ \pi/4 < \phi < \pi/2 & \textit{when } \gamma < 0,\ \mu_3 < -1 \\ \pi/2 < \phi < 3\pi/4 & \textit{when } \gamma > 0,\ \mu_3 < -1 \\ 3\pi/4 < \phi < \pi & \textit{when } \gamma < 0,\ \mu_3 > 1 \end{array}\right\} \tag{4.11.9}$$

The stresses for the steady-state are then given by

$$\left.\begin{aligned} t_{rr} &= -p + \lambda_1\sin^2\phi + A\gamma \\ t_{\theta\theta} &= -p + \lambda_1\cos^2\phi + B\gamma \\ t_{zz} &= -p\ ,\ t_{\theta z} = 0\ ,\ t_{zr} = 0 \\ t_{r\theta} &= C + D\gamma \end{aligned}\right\} \tag{4.11.10}$$

where

$$\left.\begin{aligned} 2A &= (\lambda^2\sin^2\phi + 2\lambda_4)\sin 2\phi \\ 2B &= (\lambda_2\cos^2\phi + 2\lambda_4)\sin 2\phi \\ 2C &= \lambda_1 \sin 2\phi \\ 4D &= 4(\lambda_3 + \lambda_4) + \lambda_2 \sin^2 2\phi \end{aligned}\right\} \tag{4.11.11}$$

Since a reversal of shear should not change the normal stresses, but should only reverse the corresponding shear stress, Eqs. (4.11.10) show that the constants A, B and C reverse signs when γ does. Therefore special care must be taken to consider regions where $\gamma = 0$. For the present, $\gamma \neq 0$ will be assumed to hold in the annulus. Equation (4.11.5) b may then be integrated to give

$$t_{r\theta} = \frac{T}{2\pi r^2} \tag{4.11.12}$$

where T is the torque per unit height. A substitution from this in Eq. (4.11.10) f gives

$$\gamma = \frac{T}{2\pi D r^2} - \frac{C}{D} \tag{4.11.13}$$

Since stresses are expected to do non-negative work, $\gamma t_{r\theta} \geqslant 0$, which requires that $\gamma(C + D\gamma) \geqslant 0$, so that C and γ must be of the same sign and $D > 0$. For this to be possible, Eqs. (4.11.9) a, d and (4.11.11) show that the following conditions must be satisfied

$$\left.\begin{aligned} &\lambda_1 \geqslant 0,\ \lambda_2(\mu_3^2 - 1) + 4\mu_3^2(\lambda_3 + \lambda_4) \geqslant 0,\ \textit{when}\ \mu_3 > 1 \\ &\lambda_1 \leqslant 0,\ \lambda_2(\mu_3^2 - 1) + 4\mu_3^2(\lambda_3 + \lambda_4) \geqslant 0,\ \textit{when}\ \mu_3 < -1 \end{aligned}\right\} \tag{4.11.14}$$

Then Eq. (4.11.13) shows that T and γ must be of the same sign. Equation (4.11.13) shows that γ will not be zero anywhere in the annulus provided that the torque per unit height T satisfies

$$\frac{1}{2\pi} |T| > |C| R_2^2 \tag{4.11.15}$$

Since reversing the sign of T only amounts to changing the direction of the torque, in which case the solution is easily determined, T will be assumed to be positive.

An integration of Eq. (4.11.5) a, with the help of Eqs. (4.11.10) a and b, and the expression for γ given in Eq. (4.11.13), then gives

$$p = \frac{(A+B)T}{4\pi D r^2} - \left[\lambda \cos 2\phi + (A - B)\frac{C}{D}\right] \ln r + \rho \int \frac{v^2}{r}\, dr + E \tag{4.11.16}$$

where E is an arbitrary constant in which some other constants, that come up during integration, have been absorbed. Also, an integration of Eq. (4.11.13) gives

$$v = Pr - \frac{C}{D} r \ln r - \frac{T}{4\pi Dr} \tag{4.11.17}$$

where P is an arbitrary constant. An application of the boundary conditions, given in Eq. (4.11.2), to Eq. (4.11.17) gives

$$\frac{T}{4\pi D} = \frac{R_1^2 R_2^2}{R_2^2 - R_1^2} \left[(\Omega_2 - \Omega_1) + \frac{C}{D} \ln(R_2/R_1)\right] \tag{4.11.18}$$

$$P = \frac{1}{R_2^2 - R_1^2} \left[(R_2^2 \Omega_2 - R_1^2 \Omega_1) + \frac{C}{D} (R_2^2 \ln R_2 - R_1^2 \ln R_1)\right] \tag{4.11.19}$$

Thus, Eq. (4.11.17), with P given by Eq. (4.11.19), gives an acceptable solution as long as the condition in Eq. (4.11.15) holds. The torque for this case is given by Eq. (4.11.18).

In the case when the condition $|T|/2\pi > |C| R_2^2$ is violated, there are two possibilities

$$\frac{T}{2\pi} > C R_1^2 \tag{4.11.20}$$

or

$$\frac{T}{2\pi} < C R_1^2 \tag{4.11.21}$$

where T is taken to be positive. If Eq. (4.11.20) holds, then the above solution can be adapted for $R_1 \leqslant r \leqslant \bar{R}$, where $\bar{R}^2 = T/2\pi C$, and a rigid rotation may be assumed for $\bar{R} \leqslant R \leqslant R_2$, giving a physically acceptable solution for all r. On the other hand, if Eq. (4.11.21) holds, then a rigid rotation is assumed for the entire annulus $R_1 \leqslant r \leqslant R_2$. This Bingham-fluid-like behavior of anisotropic fluids has also been predicted, in Section (4.10), for steady Poiseuille flow through circular pipes.

A discussion on the stresses in the rigidly rotating annulus, when either Eq. (4.11.20) or (4.11.21) holds, is given in Ref. [7], from which the material for this section has been abstracted.

Equations (4.11.3) and (4.11.6) show that even though the macroscopic velocity field $v = v(r)$ is steady, the dependence of $\mathbf{n}$ on time gives rise to a time varying stress distribution that may be interpreted as a stress relaxation phenomenon.

4.12 Concluding Remarks

In Ericksen's theory, which has been presented in this chapter, the stress tensor is not symmetric, in general, even when couple stresses and body moments are absent. Equation (4.4.15) shows that the stress depends on the body moment. This dependence is objectionable on physical grounds. However, this difficulty does not arise in the absence of body moments. Note that the same difficulty also arises in the theory of couple stresses in fluids given in Chapter 3.

However, solutions have only been obtained for the case in which the molecular inertia is negligible, so that the stress tensor is symmetric. Only linear constitutive equations have been considered. Solutions have only been discussed for the isothermal flows of incompressible fluids. The material in this chapter gives the simplest treatment of the effects of microstructure in the absence of couple stresses.

Section (4.8) showed that microstructure gives rise to an effect that may be looked upon as a form of stress relaxation. Thus, even when the macroscopic velocity field is steady, the stresses can vary with time. The analysis given in Section (4.10) shows that such fluids also exhibit Bingham-fluid-like behavior. Such fluids are capable of exhibiting a host of other phenomena. Their analyses are described in the papers given in the list of references.

The ideas of Ericksen have been successfully extended by Leslie [Refs. 27, 30, 36]. Reference [8] gives a review of the applications of such theories to liquid crystals up to 1967. A review of these theories, their applications, and their relation to more general theories, is given in Refs. [9] and [10].

The theories of anisotropic fluids may now be considered to be special cases of the more general theories of Eringen, in which both couple stresses and microstructure are taken into account.

4.13 References

The bulk of the results of this chapter are based on the material contained in Refs. [1] to [7].

A review of the continuum theory of liquid crystals up to 1967 is given in Ref. [8]. A review of the relationship of the theory of anisotropic fluids to other nonclassical theories and some applications, are given respectively in Refs. [9] and [10].

References [11] and [12] describe important early work on the theory of liquid crystals.

References [13] to [15] contain material that is pertinent to the material discussed in this chapter.

The remaining references, arranged in chronological order, pertain to different aspects of the theory of anisotropic fluids.

1. Ericksen, J. L. (1960). Transversely Isotropic Fluids, *Kolloid-Z.* **173**, 117-122.

2. Ericksen, J. L. (1960). Anisotropic Fluids, *Arch. Ration. Mech. Anal.* **4**, 231-237.
3. Ericksen, J. L. (1960). Theory of Anisotropic Fluids, *Trans. Soc. Rheol.* **4**, 29-39.
4. Ericksen, J. L. (1961). Conservation Laws of Liquid Crystals, *Trans. Soc. Rheol.* **5**, 23-34.
5. Ericksen, J. L. (1961). Poiseuille Flow of Certain Anisotropic Fluids, *Arch. Ration. Mech. Anal.* **8**, 1-8.
6. Ericksen, J. L. (1962). Orientation Induced by Flow, *Trans. Soc. Rheol.* **6**, 275-291.
7. Verma, P. D. S. (1962). Couette Flow of Certain Anisotropic Fluids, *Arch. Ration. Mech. Anal.* **10**, 101-107.
8. Ericksen, J. L. (1967). Continuum Theory of Liquid Crystals, *Appl. Mech. Rev.* **20**, 1029-1032.
9. Ariman, T., Turk, M. A., and Sylvester, N. D. (1973). Microcontinuum Fluid Mechanics — A Review, *Int. J. Eng. Sci.* **11**, 905-930.
10. Ariman, T., Turk, M. A., and Sylvester, N. D. (1974). Review Article: Applications of Microcontinuum Fluid Mechanics, *Int. J. Eng. Sci.* **12**, 273-293.
11. Oseen, C. W. (1933). The Theory of Liquid Crystals, *Trans. Faraday Soc.* **29**, 883-899.
12. Frank, F. C. (1958). On the Theory of Liquid Crystals, *Discuss. Faraday Soc.* **25**, 19-28.
13. Jeffery, G. B. (1922). The Motion of Ellipsoidal Particles Immersed in a Viscous Fluid, *Proc. R. Soc. London Series A* **102**, 161-179.
14. Oldroyd, J. G. (1947). A Rational Formulation of the Equations of Plastic Flow for a Bingham Solid, *Proc. Cambridge Philos. Soc.* **43**, 100-105.
15. Smith, G. F., and Rivlin, R. S. (1957). The Anisotropic Tensors, *Q. Appl. Math.* **15**, 308-314.
16. Ericksen, J. L. (1960). A Vorticity Effect in Anisotropic Fluids, *J. Polym. Sci.* **47**, 327-331.
17. Hand, G. L. (1961). A Theory of Dilute Suspensions, *Arch. Ration. Mech. Anal.* **7**, 81-86.
18. Ericksen, J. L. (1962). Kinematics of Macromolocules, *Arch. Ration. Mech. Anal.* **9**, 1-8.
19. Ericksen, J. L. (1962). Hydrostatic Theory of Liquid Crystals, *Arch. Ration. Mech. Anal.* **9**, 371-378.

20. Ericksen, J. L. (1962). Nilpotent Energies in Liquid Crystal Theory, *Arch. Ration. Mech. Anal.* **10**, 189-196.

21. Ericksen, J. L. (1963). Singular Surfaces in Anisotropic Fluids, *Int. J. Eng. Sci.* **1**, 157-161.

22. Green, A. E. (1964). A Continuum Theory of Anisotropic Fluids, *Proc. Cambridge Philos. Soc.* **60**, 123-128.

23. Leslie, F. M. (1964). The Stability of Certain Anisotropic Fluids, *Proc. Cambridge Philos. Soc.* **60**, 949-955.

24. Coleman, B. D. (1965). Simple Liquid Crystals, *Arch. Ration. Mech. Anal.* **20**, 41-58.

25. Kaloni, P. N. (1965). On Certain Steady Flows of Anisotropic Fluids, *Int. J. Eng. Sci.* **3**, 515-532.

26. Kaloni, P. N. (1965). On the Flow of Certain Orientable Fluids Between Two Coaxial Cones, *J. Fluid Mech.* **23**, 511-520.

27. Allen, S. J., and DeSilva, C. N. (1966). A Theory of Transversely Isotropic Fluids, *J. Fluid Mech.* **24**, 801-821.

28. Denn, M. M., and Metzner, A. B. (1966). Elementary Flows of Anisotropic Fluids, *Trans. Soc. Rheol.* **10**, 215-227.

29. Ericksen, J. L. (1966). Inequalities in Liquid Crystal Theory, *Phys. Fluids* **9**, 1205-1207.

30. Ericksen, J. L. (1966). Some Magnetohydrodynamic Effects in Liquid Crystals, *Arch. Ration. Mech. Anal.* **23**, 266-275.

31. Kaloni, P. N. (1966). Certain Periodic Flows of Anisotropic Fluids, *Phys. Fluids* **9**, 1316-1321.

32. Leslie, F. M. (1966). Some Constitutive Equations for Anisotropic Fluids, *Q. J. Mech. Appl. Math.* **19**, 357-370.

33. Allen, S. J., DeSilva, C. N., and Kline, K. A. (1967). Theory of Simple Deformable Directed Fluids, *Phys. Fluids* **10**, 2551-2555.

34. Ericksen, J. L. (1967). Twisting of Liquid Crystals, *J. Fluid Mech.* **27**, 59-64.

35. Allen, S. J., and Kline, K. A. (1968). Rectilinear Shear Flow of Fluids with Interacting Deformable Substructure, *Z. Angew. Math. Phys.* **19**, 425-433.

36. Dafermos, C. M. (1968). Stability of Orientation Patterns of Liquid Crystals Subject to Magnetic Fields, *SIAM J. Appl. Math.* **16**, 1305-1318.

37. DeSilva, C. N., and Kline, K. A. (1968). Nonlinear Constitutive Equa-

tions for Directed Viscoelastic Materials with Memory, *Z. Angew. Math. Phys.* **19**, 128-139.

38. Ericksen, J. L. (1968). Twist Waves in Liquid Crystals, *Q. J. Mech. Appl. Math.* **21**, 463-465.

39. Ericksen, J. L. (1968). Twisting of Partially Oriented Liquid Crystals, *Q. Appl. Math.* **25**, 474-479.

40. Kirwan, A. D., Jr. (1968). Constitutive Equations for a Fluid Containing Non-Rigid Structures, *Phys. Fluids* **11**, 1440-1446.

41. Kline, K. A., and Allen, S. J. (1968). On Continuum Theories of Suspensions of Deformable Particles, *Z. Angew. Math. Phys.* **19**, 898-905.

42. Kline, K. A., and Allen, S. J. (1968). Heat Conduction in Fluids with Substructure, *Z. Angew. Math. Mech.* **48**, 435-443.

43. Leslie, F. M. (1968). Some Constitutive Equations for Liquid Crystals, *Arch. Ration. Mech. Anal.* **28**, 265-283.

44. Leslie, F. M. (1968). Some Thermal Effects in Cholesteric Liquid Crystals, *Proc. R. Soc. London Series A* **307**, 359-372.

45. Ericksen, J. L. (1969). A Boundary Layer Effect in Viscometry of Liquid Crystals, *Trans. Soc. Rheol.* **13**, 9-15.

46. Ericksen, J. L. (1969). Continuum Theory of Liquid Crystals of the Nematic Type, *Mol. Cryst. Liq. Cryst.* **7**, 153-164.

47. Ericksen, J. L. (1969). Twisting of Liquid Crystals by Magnetic Fields, *Z. Angew. Math. Phys.* **20**, 383-388.

48. Kaloni, P. N., and DeSilva, C. N. (1969). Oriented Fluids and the Rheology of Suspensions, *Phys. Fluids* **12**, 994-999.

49. Kaloni, P. N., and DeSilva, C. N. (1969). Elongational Flow of a Dilute Suspension of Viscoelastic Spheres, *Phys. Fluids* **12**, 2437-2438.

50. Kaloni, P. N., and DeSilva, C. N. (1969). On the Rheology of Dilute Suspensions of Rigid Particles, *Trans. Soc. Rheol.* **14**, 319-334.

51. Kirwan, A. D., and Newman, N. (1969). Simple Flow of a Fluid Containing Deformable Structures, *Int. J. Eng. Sci.* **7**, 1067-1078.

52. Kline, K. A., and Allen, S. J. (1969). Nonsteady Flows of Fluids with Microstructure, *Phys. Fluids* **13**, 263-270.

53. Kline, K. A., and Allen, S. J. (1969). The Relationship of Pressure Gradient to Blood Velocity Based on a Continuum Theory of Blood, *J. Biomech.* **2**, 313-318.

54. Leslie, F. M. (1969). Continuum Theory of Cholesteric Liquid Crystals, *Mol. Cryst. Liq. Cryst.* **7**, 407-420.

55. Allen, S. J., and Kline, K. A. (1970). Fluid Suspensions; Flow Near an Oscillating Plate, *Trans. Soc. Rheol.* **14**, 39-45.

56. Atkin, R. J. (1970). Poiseuille Flow of Liquid Crystals of the Nematic Type, *Arch. Ration. Mech. Analy.* **38**, 224-240.

57. Atkin, R. J., and Leslie, F. M. (1970). Couette Flow of Nematic Liquid Crystals, *Q. J. Mech. Appl. Math.* **23**, 53-524.

58. Currie, P. K. (1970). Couette Flow of a Nematic Liquid Crystal in the Presence of a Magnetic Field, *Arch. Ration. Mech. Anal.* **37**, 222-242.

59. Dafermos, C. M. (1970). Disinclinations in Liquid Crystals, *Q. J. Mech. Appl. Math.* **23**, S49-S64.

60. Kaloni, P. N., and DeSilva, C. N. (1970). A Theory of Oriented Fluids, *Phys. Fluids* **13**, 1708-1716.

61. Kaloni, P. N., and DeSilva, C. N. (1970). On the Rheology of Dilute Suspensions of Rigid Particles, *Trans. Soc. Rheol.* **14**, 319-334.

62. Kirwan, A. D., Jr., and Chang, M. S. (1970). Cylindrical Flow of a Fluid Containing Deformable Structures, *Int. J. Eng. Sci.* **8**, 731-742.

63. Kline, K. A., and Allen, S. J. (1970). Nonsteady Flows of Fluids with Microstructure, *Phys. Fluids* **13**, 263-270.

64. Allen, S. J., and Kline, K. A. (1971). A Thermodynamical Theory of Fluid Suspensions, *Phys. Fluids* **14**, 1863-1969.

65. Allen, S. J., Kline, K. A., and Ling, C. C. (1971). Kinematics of Deformable Suspensions, *Trans. Soc. Rheol.* **15**, 177-188.

66. Lee, J. D., and Eringen, A. C. (1971). Wave Propagation in Nematic Liquid Crystals, *J. Chem. Phys.* **54**, 5027-5034.

67. Lee, J. D., and Eringen, A. C. (1971). Alignment of Nematic Liquid Crystals, *J. Chem. Phys.* **55**, 4504-4508.

68. Lee, J. D., and Eringen, A. C. (1971). Boundary Effects of Orientation of Nematic Liquid Crystals, *J. Chem. Phys.* **55**, 4509-4512.

69. Gordon, R. J., and Schowalter, W. R. (1972). Anisotropic Fluid Theory: A Different Approach to the Dumbell Theory of Dilute Polymer Solutions, *Trans. Soc. Rheol.* **16**, 79-97.

70. Kline, K. A., and Carmi, S. (1972). On the Stability of Motions of a Dilute Suspension of Rigid Spherical Particles: Universal Stability, *Bull. Acad. Pol. Sci. Ser. Sci. Tech.* **20**, 693-699.

71. Allen, S. J., Kline, K. A., and Ling, C. C. (1973). Transient Shear Flow of Fluids with Deformable Microstructures, *Acta Mech.* **18**, 1-20.

72. Currie, P. K. (1973). Static Solutions with Phase Change for a Nematic Liquid Crystal in a Pipe, *Trans. Soc. Rheol.* **17**, 197-208.

73. Kline, K. A. (1973). Polymers as Structural Continua: Superposed Oscillatory Shear Flows, *Trans. Soc. Rheol.* **17**, 525-536.

74. Barratt, P. J. (1974). A Continuum Model for Disinclination Lines in Nematic Liquid Crystals, *Q. J. Mech. Appl. Math.* **27**, 505-522.

75. Barratt, P. J., and Jenkins, J. T. (1974). Interfacial Effects in the Static Theory of Nematic Liquid Crystals, *Q. J. Mech. Appl. Math.* **27**, 11-127.

76. Ericksen, J. L. (1974). Liquid Crystals and Cosserat Surfaces, *Q. J. Mech. Appl. Math.* **27**, 213-219.

77. Levitskii, S. P., and Listrov, A. T. (1974). Stability of Flow of a Liquid Crystal Layer on an Inclined Plane, *J. Appl. Math. Mech. (USSR)* **38**, 978-987.

78. Ericksen, J. L. (1976). On Equations of Motion for Liquid Crystals, *Q. J. Mech. Appl. Math.* **29**, 203-208.

79. Leslie, F. M. (1976). Analysis of a Flow Instability in Nematic Liquid Crystals, *J. Phys. D* **9**, 925-937.

80. Sandberg, T. K., and Kline, K. A. (1976). Structured Fluid Theory, *Acta Mech.* **24**, 141-155.

81. Shahinpoor, M. (1976). On the Stress Tensor in Nematic Liquid Crystals, *Rheol. Acta* **15**, 99-103.

82. Suyazov, V. M. (1976). Hydrodynamics of a Fluid with a Deformable Microstructure, *Sov. Appl. Mech.* **11**, 1108-1115.

83. Dmitriev, N. M. (1977). Anisotropy Effects in Polymer Solution Flow, *Moscow Univ. Mech. Bull.* **32**, 66-70.

84. Dubois-Violette, E., and Manneville, P. (1978). Stability of Couette Flow in Nematic Liquid Crystals, *J. Fluid Mech.* **89**, 273-303.

85. Duffy, B. R. (1978). Flow of a Liquid with an Anisotropic Viscosity Tensor, *J. Non-Newtonian Fluid Mech.* **4**, 177-193.

86. Currie, P. K. (1979). Apparent Viscosity During Viscometric Flow of Nematic Liquid Crystals, *J. Phys. (Paris)* **40**, 501-505.

87. Currie, P. K., and MacSithigh, G. P. (1979). Stability and Dissipation of Solutions for Shearing Flow of Nematic Liquid Crystals, *Q. J. Mech. Appl. Math.* **32**, 499-511.

88. Manneville, P. (1979). Theoretical Analysis of Poiseuille Flow Instabilities in Nematics, *J. Phys. (Paris)* **40**, 713-724.

89. Duffy, B. R. (1980). Flow of a Liquid with an Anisotropic Viscosity Tensor: Inertial Effects, *J. Non-Newtonian Fluid Mech.* **7**, 107-120.

90. Duffy, B. R. (1980). Flow of a Liquid with an Anisotropic Viscosity

Tensor: Some Axisymmetric Flows, *J. Non-Newtonian Fluid Mech.* **7**, 359-367.

91. Eringen, A. C. (1980). Theory of Anisotropic Micropolar Fluids, *Int. J. Eng. Sci.* **18**, 5-17.

92. Barratt, P. J. (1981). Calculation of Shear Rate Thresholds for Homogeneous and Roll Type Instabilities in Nematics Subjected to Simple Shear Flow, *J. Phys. D* **14**, 1831-1846.

93. Demiray, H. (1981). Constitutive Equation for the Nematic Liquid Crystals, *J. Tech. Phys.* **22**, 217-232.

94. Duffy, B. R. (1981). Flow of a Liquid with an Anisotropic Viscosity Tensor: A Skew Initial Orientation, *J. Non-Newtonian Fluid Mech.* **8**, 213-222.

95. Duffy, B. R. (1981). Flow of a Liquid with an Anisotropic Viscosity Tensor: Elongational Flow, *J. Non-Newtonian Fluid Mech.* **9**, 1-12.

96. Leslie, F. M. (1981). Viscometry of Nematic Liquid Crystals, *Mol. Crys. Liq. Crys.* **63**, 111-127.

97. Petrov, M., and Durand, G. (1981). On the Origin of Electrohydrodynamic Instabilities in Smectic C Liquid Crystals, *J. Phys. (Paris) Lett.* **42**, 519-522.

98. Wissbrun, K. F. (1981). Rheology of Rod-Like Polymers in the Liquid Crystalline State, *J. Rheol.* **25**, 619-662.

99. Barratt, P. J., and Zuniga, I. (1982). Theoretical Investigation of the Pieranski-Guyon Instability in Couette Flow of Nematic Liquid Crystals, *J. Non-Newtonian Fluid Mech.* **11**, 23-36.

100. Carlsson, T. (1982). Possibility of the Existence of a Positive Leslie Viscosity α_2. Proposed Flow Behavior of Disk Like Nematic Liquid Crystals, *Mol. Cryst. Liq. Cryst.* **89**, 57-66.

101. Volkov, V. S. (1982). Slow Motion of a Sphere Through an Anisotropic, Viscoelastic Fluid, *J. Appl. Math. Mech. (USSR)* **46**, 185-189.

CHAPTER 5

Micro Fluids

5.1 Introduction

This chapter discusses a class of fluids with microstructure, termed micromorphic fluids by A. C. Eringen. Two important micromorphic fluids are micro fluids and a subset called micropolar fluids.

A special type of microstructure was introduced, in a heuristic way, by J. L. Ericksen, who showed that microstructure can account for various phenomena such as stress relaxation and plug-type flow. An account of this theory has been given in Chapter 4.

This chapter discusses the work of A. C. Eringen, who has developed a class of theories in which both the effects of couple stresses and microstructure are simultaneously taken into account in a systematic manner.

5.2 Description of Micromotion

In continuum theories the mass is assumed to be a continuous measure, so that a continuous mass density ρ exists in a volume element dV that is infinitesimally small. However, this continuum concept of density breaks down when the volume compares with the cube of the mean free path, say ΔV^*. Thus the mathematical idealization is meaningful only when dV models ΔV such that $\Delta V > \Delta V^*$. The theory of micromorphic fluids attempts to account for the microstructure that exists for $\Delta V < \Delta V^*$ by using a continuum description. The limiting volume ΔV^* could also be interpreted as a volume, containing microstructures such as suspended particles or even turbulent structures, for which the effects of microstructure cannot be accounted for by a continuum distribution.

In order to account for the microstructure that exists in $\Delta V < \Delta V^*$, the deformations of the material within ΔV have to be considered. The macromass element dM is assumed to contain continuous mass distributions such that the total macromass dM is the sum of all the masses in dV. Thus a continuous mass distribution is assumed to exist at each point of the macroelement dV, such that the sum of the local masses over dV gives the total mass dM, and implies that the continuum theory is valid at each point of a macroelement dV. Statistical averages then have to be taken to obtain the microdeformation theory of continuous media.

Consider a body of fluid that occupies the region of space $V + S$ in the undisturbed state at time t_0, and the region $\overline{V} + \overline{S}$ in the deformed state at time t. Let dV be a macrovolume element in the undisturbed fluid, having within it a continuous distribution of microvolume elements dV', as shown in Fig. 5.2.1. dS and dS' are then the surface elements of the macro and

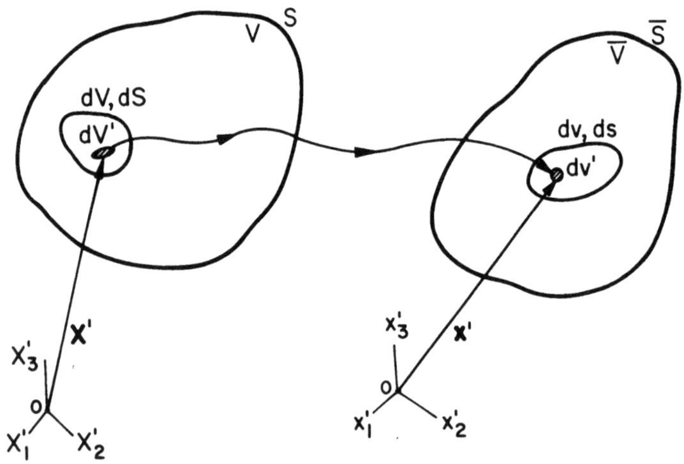

Fig. 5.2.1 Macroelements and microelements of a body at two different times.

micro volume elements, respectively. On deformation, these macro and micro elements will become dv and dv', as shown. The macroelements are constituted from the microelements such that $dV = \int_{dV} dV'$ and $dv = \int_{dv} dv'$.

A material point $\mathbf{X}'$ having Cartesian coordinates X'_K in the undisturbed body $V + S$ at time t_0, occupies the spatial point $\mathbf{x}'$, with Cartesian coordinates x'_k, in the deformed body $\bar{V} + \bar{S}$ at time t. The motion and the inverse motion are given by the one parameter one-to-one mappings

$$\mathbf{x}' = \mathbf{x}'(\mathbf{X}', t) \quad , \quad \mathbf{X}' = \mathbf{X}'(\mathbf{x}', t) \tag{5.2.1}$$

In this description $\mathbf{X}'$ may be measured relative to different Cartesian frames.

Quantities associated with the material point $\mathbf{X}'$ will be denoted by primed upper case letters and those associated with the spatial point $\mathbf{x}'$ by primed lower case letters. For example dV' and dv' denote the microvolume elements centered at $\mathbf{X}'$ and $\mathbf{x}'$, respectively.

Let P' and ρ' be the mass densities at $\mathbf{X}'$ and $\mathbf{x}'$, respectively. Then the macro or average mass densities, P and ρ, in the macrovolume elements dV and dv, respectively, are defined by

$$\mathrm{P}\, dV = \int_{dV} \mathrm{P}'\, dV' \quad , \quad \rho\, dv = \int_{dv} \rho'\, dv' \tag{5.2.2}$$

Let **X** be the center of mass of the undeformed macroelement, in the configuration at time t_0, then

$$\mathrm{P}\mathbf{X}\ dV = \int_{dV} \mathrm{P}'\mathbf{X}'\ dV' \tag{5.2.3}$$

Because of the motion, the material point **X** is carried into a spatial point **x** at time t, so that

$$\mathbf{x} = \mathbf{x}(\mathbf{X},t) \quad , \quad \mathbf{X} = \mathbf{X}(\mathbf{x},t) \tag{5.2.4}$$

Let

$$\mathbf{X}' = \mathbf{X} + \boldsymbol{\Xi} \quad , \quad \mathbf{x}' = \mathbf{x} + \boldsymbol{\xi} \tag{5.2.5}$$

Then Ξ_K are the coordinates of the material point $\mathbf{X}'$ in dV relative to the center of mass **X** of the macroelement dV. Use of Eqs. (5.2.3), (5.2.5) and (5.2.2) then gives

$$\mathrm{P}\mathbf{X}\ dV = \int_{dV} \mathrm{P}'\mathbf{X}\ dV' = \int_{dV} \mathrm{P}'(\mathbf{X} + \boldsymbol{\Xi})\ dV'$$

$$= \mathbf{X}\int_{dV} \mathrm{P}'\ dV' + \int_{dV} \mathrm{P}'\boldsymbol{\Xi}\ dV'$$

$$= \mathrm{P}\mathbf{X}\ dV + \int_{dV} \mathrm{P}'\boldsymbol{\Xi}\ dV'$$

so that

$$\int_{dV} \mathrm{P}'\boldsymbol{\Xi}\ dV' = 0 \tag{5.2.6}$$

which is in agreement with the assumption that $\boldsymbol{\Xi}$ is the position of a point in dV relative to its center of mass.

The geometry of the motion is shown in Fig. 5.2.2. The center of mass

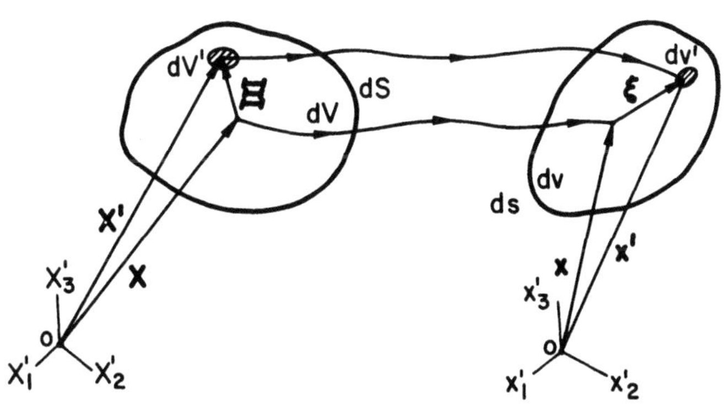

Fig. 5.2.2 Geometry of microelements within macroelements.

$\mathbf{X}$ of dV goes into the point $\mathbf{x}$, and the point $\mathbf{X}' = \mathbf{X} + \boldsymbol{\Xi}$ goes into $\mathbf{x}' = \mathbf{x} + \boldsymbol{\xi}$. $\mathbf{x}$ will now be shown to be the center of mass of dv as long as $|\boldsymbol{\Xi}|$ is small.

Equations (5.2.1) and (5.2.5) give

$$\mathbf{x}' = \mathbf{x}'(\mathbf{X}', t) = \mathbf{x}'(\mathbf{X} + \boldsymbol{\Xi}, t) = \mathbf{x}(\mathbf{X}, t) + \boldsymbol{\xi} \tag{5.2.7}$$

so that $\boldsymbol{\xi}$ is a function of $\mathbf{X}$, $\boldsymbol{\Xi}$ and t. That is

$$\boldsymbol{\xi} = \boldsymbol{\xi}(\mathbf{X}, \boldsymbol{\Xi}, t) \tag{5.2.8}$$

An expansion of $\boldsymbol{\xi}$ around $\boldsymbol{\Xi} = 0$ then gives

$$\boldsymbol{\xi} = \boldsymbol{\xi}(\mathbf{X}, 0, t) + \frac{\partial \boldsymbol{\xi}}{\partial \Xi_K}\Big|_{\boldsymbol{\Xi}=0} \Xi_K + \cdots \tag{5.2.9}$$

With $\boldsymbol{\Xi} = 0$, from Eq. (5.2.7) $\mathbf{x}(\mathbf{X}, t) = \mathbf{x}(\mathbf{X}, t) + \boldsymbol{\xi}(\mathbf{X}, 0, t)$, so that $\boldsymbol{\xi}(\mathbf{X}, 0, t) = 0$. Thus, for sufficiently small $|\boldsymbol{\Xi}|$, if only linear terms in $\boldsymbol{\Xi}$ are retained in Eq. (5.2.9),

$$\boldsymbol{\xi}(\mathbf{X}, \boldsymbol{\Xi}, t) = \frac{\partial \boldsymbol{\xi}}{\partial \Xi_K}\Big|_{\boldsymbol{\Xi}=0} \Xi_K = \boldsymbol{\chi}_K(\mathbf{X}, t) \Xi_K \tag{5.2.10}$$

where

$$\boldsymbol{\chi}_K(\mathbf{X}, t) = \frac{\partial \boldsymbol{\xi}}{\partial \Xi_K}\Big|_{\boldsymbol{\Xi}=0} \Xi_K \tag{5.2.11}$$

Assuming that the mass of a microelement is conserved,

$$\mathrm{P}' \, dV' = \rho' \, dv' \tag{5.2.12}$$

so that from Eq. (5.2.2)

$$\mathrm{P} \, dV = \rho \, dv \tag{5.2.13}$$

which is essentially the conservation of mass of a macroelement.

Subject to the approximation $\boldsymbol{\xi} = \boldsymbol{\chi}_K(\mathbf{X}, t)\Xi_K$, the motion $\mathbf{x}' = \mathbf{x}'(\mathbf{X} + \boldsymbol{\Xi}, t) = \mathbf{x}(\mathbf{X}, t) + \boldsymbol{\xi}$ carries the center of mass of dV into the center of mass of dv. For,

$$\int_{dv} \rho' \mathbf{x}' \, d\mathbf{v}' = \int_{dv} \rho'(\mathbf{x} + \boldsymbol{\chi}_K \Xi_K) \, dv' = \mathbf{x} \int_{dv} \rho' \, dv' + \boldsymbol{\chi}_K \int_{dv} \Xi_K \rho' \, dv'$$
$$= \mathbf{x} \rho \, dv + \boldsymbol{\chi}_K \int_{dV} \mathrm{P}' \Xi_K \, dV' = \rho \mathbf{x} \, dv$$

since $\int_{dV} P'\Xi_K \, dV' = 0$, thereby showing that $\mathbf{x}$, which is the image of $\mathbf{X}$ due to the motion, is the center of mass of dv.

In $\boldsymbol{\xi} = \boldsymbol{\chi}_K \Xi_K$, $\boldsymbol{\chi}_K$, is a vector for each value of K. Let $(\boldsymbol{\chi}_K)_k = \chi_{Kk}$ be the *kth* component of $\boldsymbol{\chi}_K$. Then this result can be expressed in terms of components as

$$\xi_k = \chi_{Kk}\Xi_K \quad , \quad \chi_{Kk} = \frac{\partial \xi_k}{\partial \Xi_K}\Big|_{\Xi_k=0} \tag{5.2.14}$$

Note that $\chi_{Kk} = \chi_{Kk}(\mathbf{X},t)$ is a function of $\boldsymbol{\Xi}$ and t only, and does not depend on $\boldsymbol{\Xi}$ or $\boldsymbol{\xi}$.

Similarly, for the inverse motion,

$$\mathbf{X}' = \mathbf{X}(\mathbf{x},t) + \boldsymbol{\Xi} = \mathbf{X}(\mathbf{x},t) + \frac{\partial \boldsymbol{\Xi}}{\partial \xi_k}\xi_k \tag{5.2.15}$$

gives

$$\boldsymbol{\Xi} = \boldsymbol{\chi}_k \xi_k \quad or \quad \Xi_K = \chi_{kK}\xi_k \tag{5.2.16}$$

where $\chi_{kK} = \partial \Xi_K / \partial \xi_k |_{\xi_k=0}$. Then

$$\chi_{kK}\chi_{Kl} = \delta_{kl} \quad , \quad \chi_{Kk}\chi_{kL} = \delta_{KL} \tag{5.2.17}$$

The velocity $\mathbf{v}' = \dot{\mathbf{x}}'$ and the acceleration $\mathbf{a}' = \dot{\mathbf{v}}' = \ddot{\mathbf{x}}'$ of a material point $\mathbf{X}'$ are defined by

$$\dot{\mathbf{x}}' = \frac{d\mathbf{x}'}{dt}\Big|_{\mathbf{X}'} \quad , \quad \ddot{\mathbf{x}}' = \frac{d\dot{\mathbf{x}}'}{dt}\Big|_{\mathbf{X}'}$$

so that

$$\dot{\mathbf{x}}' = \frac{d}{dt}\mathbf{x}'|_{\mathbf{X}'} = \frac{d}{dt}[\mathbf{x}(\mathbf{X},t) + \boldsymbol{\chi}_K(\mathbf{X},t)\Xi_K]\mathbf{X}'$$

$$= \frac{d}{dt}\mathbf{x}|_{\mathbf{X}} + \frac{d}{dt}\boldsymbol{\chi}_K|_{\mathbf{X},\boldsymbol{\Xi}}\Xi_K$$

which may be written as

$$\dot{\mathbf{x}}' = \dot{\mathbf{x}} + \dot{\boldsymbol{\chi}}_K\Xi_K = \dot{\mathbf{x}} + \dot{\boldsymbol{\xi}} \tag{5.2.18}$$

where, for the unprimed quantities, a superposed dot indicates the time derivative with $\mathbf{X}$ fixed. In terms of components, this equation has the form $v'_k = v_k + \dot{\xi}_k = v_k + \dot{\chi}_{Kk}\Xi_K = v_k + \dot{\chi}_{Kk}\chi_{lK}\xi_l$ or

$$v'_k = v_k + \nu_{lk}\xi_l \quad , \quad \nu_{lk} = \chi_{lK}\dot{\chi}_{Kk} \tag{5.2.19}$$

where $v_k = v_k(\mathbf{x},t) = \dot{x}_k(\mathbf{X}(\mathbf{x},t),t)$ is the velocity of the center of mass $\mathbf{x}$ and

$$\dot{\xi}_k = \nu_{lk}\xi_l \tag{5.2.20}$$

is the peculiar velocity, that is, the velocity of the microelement relative to that of $\mathbf{x}$. Note that ν_{lk} does not depend on $\boldsymbol{\Xi}$ or $\boldsymbol{\xi}$.

Using $\mathbf{x}' = \mathbf{x} + \boldsymbol{\xi} = \mathbf{x} + \boldsymbol{\chi}_K \Xi_K$, the acceleration $\mathbf{a}' = d\mathbf{v}'/dt|_{\mathbf{X}'} = \ddot{\mathbf{x}}'$ is given by

$$\ddot{\mathbf{x}}' = \ddot{\mathbf{x}} + \ddot{\boldsymbol{\xi}} = \ddot{\mathbf{x}} + \ddot{\boldsymbol{\chi}}_K \Xi_K \tag{5.2.21}$$

which can also be written as

$$a'_k = a_k + \ddot{\chi}_{Kk}\Xi_K = a_k + \ddot{\chi}_{Kk}\chi_{lK}\xi_l \tag{5.2.22}$$

Now from $\nu_{lk} = \chi_{lK}\dot{\chi}_{Kk}$, $\chi_{Kl}\nu_{lk} = \chi_{Ll}\chi_{lK}\dot{\chi}_{Kk} = \dot{\chi}_{Lk}$. That is

$$\dot{\chi}_{Kk} = \chi_{Kl}\nu_{lk} \tag{5.2.23}$$

Then $\ddot{\chi}_{Kk} = \dot{\chi}_{Kl}\nu_{lk} + \chi_{Kl}\dot{\nu}_{lk}$, gives $\chi_{lK}\ddot{\chi}_{Kk} = \chi_{lK}\dot{\chi}_{Kr}\nu_{rk} + \chi_{lK}\chi_{Kr}\dot{\nu}_{rk}$, which reduces to

$$\chi_{lK}\ddot{\chi}_{Kk} = \nu_{lr}\nu_{rk}\dot{\nu}_{lk} \tag{5.2.24}$$

Using Eqs. (5.2.22) and (5.2.24), the acceleration is finally given by

$$a'_k = a_k + (\dot{\nu}_{lk} + \nu_{lr}\nu_{rk})\xi_l \tag{5.2.25}$$

Here a_k is the acceleration of the center of mass $\mathbf{x}$ and $\ddot{\xi}_k = (\dot{\nu}_{lk} + \nu_{lr}\nu_{rk})\xi_l$ is the acceleration of a microelement relative to the center of mass.

The center of mass is conserved as long as the assumption $\xi_k = \chi_{Kk}\Xi_K$ is valid. That is, the same material element continues to be the center of mass as the material undergoes motion. The centers of momentum and inertia will now be shown to coincide with the center of mass, that is

$$\int_{dv} \rho' \dot{\mathbf{x}}' \, dv' = \rho \dot{\mathbf{x}} \, dv \quad , \quad \int_{dv} \rho' \ddot{\mathbf{x}}' \, dv' = \rho \ddot{\mathbf{x}} \, dv \tag{5.2.26}$$

For, upon using Eq. (5.2.19),

$$\begin{aligned} \int_{dv} \rho' \dot{x}'_k \, dv' &= \int_{dv} \rho'(\dot{x}_k + \nu_{lk}\xi_l) \, dv' \\ &= \dot{x}_k \int_{dv} \rho' \, dv' + \nu_{lk} \int_{dv} \rho' \xi_l \, dv' \\ &= \dot{x}_k \rho \, dv \end{aligned}$$

since $\int_{dv} \rho' \xi_l \, dv' = 0$. Similarly, starting with Eq. (5.2.25),

$$\int_{dv} \rho' \ddot{x}_k \, dv' = \int_{dv} \rho' [a_k + \dot{\nu}_{lk} + \nu_{lr}\nu_{rk})\xi_l] dv'$$

$$= a_k \int_{dv} \rho' \, dv' + (\dot{\nu}_{lk} + \nu_{lr}\nu_{rk}) \int_{dv} \rho' \xi_l \, dv'$$

$$= a_k \rho \, dv$$

The motion at the microelement level is determined by the tensor $\mathbf{N} = (\nu_{ij})$ which determines the motion of a microelement relative to the center of mass of a macroelement. This important tensor is called the gyration tensor.

5.3 Kinematics of Deformation

In this section some aspects of deformation measures for microelements are discussed. First an expression is derived for the time rate of variation of a material element $d\xi_k$.

Differentiation of $\xi_k = \chi_{Kk} \Xi_K$ gives $d\xi_k = (d\chi_{Kk})\Xi_K + \chi_{Kk} d\Xi_K$, so that

$$\frac{D}{Dt}(d\xi_k) = \frac{D}{Dt}(d\chi_{Kk})\Xi_K + \frac{D}{Dt}(\chi_{Kk}) \, d\Xi_K \tag{5.3.1}$$

where $D/Dt \equiv d/dt|_{\mathbf{X}'}$ is the material time derivative.

Now

$$\frac{D}{Dt}(d\chi_{Kk}) = \frac{D}{Dt}(\chi_{Kk,l} dx_l) = \frac{D}{Dt}(\chi_{Kk,l}) \, dx_l + \chi_{Kk,l} \frac{D}{Dt}(dx_l) \tag{5.3.2}$$

In this equation

$$\frac{D}{Dt}(dx_l) = \frac{D}{Dt}\left(\frac{\partial x_l}{\partial X_R} dX_R\right) = \frac{d}{dt}\left(\frac{\partial x_l}{\partial X_R} dX_R\right)\Big|_{\mathbf{X}} = \frac{\partial}{\partial X_R}\left(\frac{dx_l}{dt}\Big|_{\mathbf{X}}\right) dX_R$$

$$= \frac{\partial v_l}{\partial X_R} dX_R = \frac{\partial v_l}{\partial X_R}\frac{\partial X_R}{\partial x_r} dx_r = \frac{\partial v_l}{\partial x_r} dx_r$$

that is

$$\frac{D}{Dt}(dx_l) = v_{l,r} \, dx_r \tag{5.3.3}$$

Also

$$\frac{D}{Dt}(\chi_{Kk,l}) = \frac{\partial}{\partial t}\chi_{Kk,l} + v_r\chi_{Kk,lr}$$

$$= (\frac{\partial}{\partial t}\chi_{Kk} + v_r\chi_{Kk,r})_{,l} - v_{r,l}\chi_{Kk,r}$$

so that

$$\frac{D}{Dt}(\chi_{Kk,l}) = (\frac{D}{Dt}\chi_{Kk})_{,l} - v_{r,l}\chi_{Kk,r} \tag{5.3.4}$$

Substitution from Eqs. (5.3.3) and (5.3.4) in Eq. (5.3.2) then gives

$$\frac{D}{Dt}(\chi_{Kk}) = (\frac{D}{Dt}\chi_{Kk})_{,l}dx_l - v_{r,l}\chi_{Kk,r}dx_l + v_{l,r}\chi_{Kk,l}dx_r$$

The last two terms cancel out and, since from Eq. (5.2.23) $\dot{\chi}_{Kk} = \chi_{Kl}\nu_{lk}$,

$$\frac{D}{Dt}(d\chi_{Kk}) = (\chi_{Kl}\nu_{lk})_{,r}dx_r \tag{5.3.5}$$

Using Eqs. (5.2.23) and (5.3.2), Eq. (5.3.1) can then be reduced as follows:

$$\frac{D}{Dt}(d\xi_k) = (\chi_{Kl}\nu_{lk})_{,r}dx_r\Xi_K + \chi_{Lk}\nu_{lk}d\Xi_K$$

$$= \chi_{Kl}\Xi_K\nu_{lk,r}dx_r + \nu_{lk}(\Xi_K\chi_{Kl,r}dx_r + \chi_{Kl}d\Xi_K)$$

$$= \xi_l\nu_{lk,r}dx_r + \nu_{lk}(\Xi_K\,d\chi_{Kl} + \chi_{Kl}d\Xi_K)$$

$$= \xi_l\nu_{lk,r}dx_r + \nu_{lk}d\xi_l$$

Thus

$$\frac{D}{Dt}(d\xi_k) = \nu_{lk}d\xi_l + \xi_l\nu_{lk,r}dx_r \tag{5.3.6}$$

Next consider two points $\mathbf{x}' = \mathbf{x} + \boldsymbol{\xi}$ and $\boldsymbol{\xi}' + d\mathbf{x}' = \mathbf{x} + \boldsymbol{\xi} + d\mathbf{x} + d\boldsymbol{\xi}$, distance ds' apart. Then

$$ds'^2 = dx'_r dx'_r = (dx_r + d\xi_r)(dx_r + d\xi_r)$$

$$= dx_r dx_r + 2dx_r d\xi_r + d\xi_r d\xi_r$$

so that its time derivative is

$$\frac{D}{Dt}(ds'^2) = 2dx_r \frac{D}{Dt}(dx_r) + 2\frac{D}{Dt}(dx_r)\,d\xi_r + 2dx_r \frac{D}{Dt}(d\xi_r) + 2d\xi_r \frac{D}{Dt}(d\xi_r)$$

Use of Eqs. (5.3.3) and (5.3.6) then gives

$$\begin{aligned}\frac{D}{Dt}(ds'^2) &= 2dx_r v_{r,l}dx_l + 2v_{r,l}dx_l d\xi_r + 2dx_r(\nu_{lr}d\xi_l + \nu_{lr,m}\xi_l dx_m) \\ &\qquad + 2d\xi_r(\nu_{lr}d\xi_l + \nu_{lr,m}\xi_l dx_m) \\ &= 2v_{r,l}dx_r dx_l + 2(v_{l,r} + \nu_{lr} + \nu_{nl,r}\xi_n)\,dx_r d\xi_l \\ &\qquad + 2\nu_{nr,l}\xi_n dx_r dx_l + 2\nu_{lr}d\xi_r d\xi_l \\ &= 2[v_{(k,l)} + \nu_{r(k,l)}\xi_r]\,dx_k dx_l \\ &\qquad + 2(v_{l,k} + \nu_{lk} + \nu_{rl,k}\xi_r)\,dx_k d\xi_l + 2\nu_{(kl)}d\xi_k d\xi_l\end{aligned}$$

This may finally be written as

$$\begin{aligned}\frac{D}{Dt}(ds'^2) &= 2[d_{kl} + a_{r(kl)}\xi_r]\,dx_k dx_l + 2(b_{lk} + a_{rlk}\xi_r)\,dx_k dx_l \\ &\qquad + 2[b_{(kl)} - d_{kl}]\,d\xi_k d\xi_l \qquad (5.3.7)\end{aligned}$$

where $d_{kl} = v_{(k,l)} = \frac{1}{2}(v_{l,k} + v_{k,l})$,

$$b_{lk} = \nu_{lk} + v_{k,l} \quad , \quad a_{klm} = \nu_{kl,m} \qquad (5.3.8)$$

A motion is said to be microrigid if $\frac{D}{Dt}(ds'^2) = 0$ for all dx_k and $d\xi_l$. This condition will be satisfied provided $d_{kl} = 0$, $b_{kl} = 0$ and $a_{klm} = 0$. Thus a necessary and sufficient condition for microrigid motion is that $d_{kl} = 0$, $b_{kl} = 0$ and $a_{klm} = 0$.

A tensor is said to be objective if it obeys the appropriate tensor transformation law under the time-dependent transformation

$$\bar{x}'_k = Q_{kr}x'_r + b'_k \quad , \quad Q_{kr}Q_{lr} = \delta_{kl} \qquad (5.3.9)$$

where $Q_{kr} = Q_{kr}(t)$ and $b'_k = b'_k(t)$ are functions of time only. This concept has been discussed in Section (1.7), where $v_{k,l}$, d_{kl} and w_{kl} have been shown to transform as

$$\left.\begin{aligned}\bar{v}_{k,l} &= Q_{kr}Q_{ls}v_{r,s} + \dot{Q}_{kr}Q_{lr}\\ \bar{d}_{kl} &= \bar{v}_{(l,k)} = Q_{kr}Q_{ls}d_{rs}\\ \bar{w}_{kl} &= \bar{v}_{[l,k]} = Q_{kr}Q_{ls}w_{rs} + Q_{kr}Q_{lr}\end{aligned}\right\} \tag{5.3.10}$$

Thus $v_{k,l}$ and w_{rs} are not objective while d_{kl} is an objective tensor.

That ν_{kl} is not objective, while b_{kl} and a_{klm} are objective tensors, will now be shown. Consider two motions which are related by Eq. (5.3.9), so that $\bar{x}'_k = Q_{kl}x'_l + b'_k$. Then, since $\bar{x}'_k = \bar{x}_k + \bar{\chi}_{Kk}\Xi_K$, where $\bar{\chi}_{Kk} = \partial\bar{\xi}_k/\partial\Xi_K$, and $x'_l = x_l + \chi_{Ll}\Xi_L$, it follows that $\bar{x}_k + \bar{\chi}_{Kk}\Xi_K = Q_{kl}(x_l + \chi_{Ll}\Xi_L) + b'_k$, which, on use of Eq. (5.3.9), reduces to

$$\bar{\chi}_{Kk} = Q_{kl}\chi_{Kl} \tag{5.3.11}$$

From this $\bar{\chi}_{mK}\bar{\chi}_{Kk} = \delta_{mk} = \bar{\chi}_{mK}Q_{lk}\chi_{Kl}$, so that $\bar{\chi}_{mK}\chi_{Kl} = Q_{kl}\delta_{mk} = Q_{ml}$ which, on multiplication by χ_{lL}, gives $\bar{\chi}_{mL} = Q_{ml}\chi_{lL}$, or

$$\bar{\chi}_{kK} = Q_{kl}\chi_{lK} \tag{5.3.12}$$

Use of Eqs. (5.3.11) and (5.3.12), together with the definition of $\bar{\nu}_{kl}$, gives

$$\begin{aligned}\bar{\nu}_{kl} &= \bar{\chi}_{kK}\dot{\bar{\chi}}_{Kl} = Q_{kr}\chi_{rK}(\dot{Q}_{ls}\chi_{Ks} + Q_{ls}\dot{\chi}_{Ks})\\ &= Q_{kr}\dot{Q}_{lr} + Q_{kr}Q_{ls}\nu_{rs}\end{aligned}$$

so that

$$\bar{\nu}_{kl} = Q_{kr}Q_{ls}\nu_{rs} + Q_{kr}\dot{Q}_{lr} \tag{5.3.13}$$

which shows that the gyration tensor ν_{kl} is not objective.

Addition of Eqs. (5.3.10) and (5.3.13) then gives

$$\begin{aligned}\bar{b}_{kl} &= \bar{\nu}_{kl} + \bar{v}_{k,l} = Q_{kr}Q_{ls}(\nu_{rs} + v_{r,s}) + (\dot{Q}_{kr}Q_{lr} + Q_{kr}\dot{Q}_{lr})\\ &= Q_{kr}Q_{ls}b_{rs}\end{aligned}$$

thereby showing that b_{kl} is an objective tensor.

Now from Eq. (5.3.13)

$$\begin{aligned}\bar{\nu}_{kl,m} &= \frac{\partial}{\partial\bar{x}'_m}(Q_{kr}Q_{ls}\nu_{rs} + Q_{kr}\dot{Q}_{lr}) = Q_{kr}Q_{ls}\nu_{rs,t}\frac{\partial x'_t}{\partial\bar{x}'_m}\\ &= Q_{kr}Q_{ls}Q_{mt}\nu_{rs,t}\end{aligned} \tag{5.3.14}$$

since $Q_{kr}\dot{Q}_{lr}$ is only a function of t. Then, from the definition of a_{klm} given in Eq. (5.3.8), a_{klm} is a third order objective tensor.

The main result of this section is that the deformation of material elements is governed by the three objective tensors $\mathbf{D} = (d_{ij})$, $\mathbf{B} = (b_{ij})$ and $\mathbf{A} = (a_{ijk})$. The tensors $\mathbf{B}$ and $\mathbf{A}$ are referred to as the second and third order microdeformation rate tensors, respectively.

Having established the kinematics, the next step is to consider the balance laws for material with microstructure.

5.4 Conservation of Mass

Section (5.2) assumed that the mass of a microelement is conserved. This requires that $\mathrm{P}'dV' = \rho' dv'$, which at the macroelement level implies that $\mathrm{P}dV = \rho\, dv$. Thus, mass is conserved at the macroelement level. The conservation of mass at the macro level is then given by the condition $D(\rho\, dv)/Dt = 0$ which, as is well-known, leads to the continuity equation

$$\dot{\rho} + \rho\, v_{k,k} = 0 \tag{5.4.1}$$

5.5 Balance of Momenta

In this theory, the local balance of linear and angular momenta at a point $\mathbf{x}'$ of the deformed microelement are assumed to be governed by Cauchy's laws of motion for a nonpolar material, that is by

$$t'_{kl,k} + \rho'(f'_l - a'_l) = 0 \tag{5.5.1}$$

$$t'_{kl} = t'_{lk} \tag{5.5.2}$$

where t'_{kl} is the stress tensor and f'_l the body force at the point $\mathbf{x}'$.

As a step towards obtaining the local balance of momenta at the center of mass, $\mathbf{x}$, of the microelement $dv + ds$, an integration of Eq. (5.5.1) over the material volume $\overline{V}$, after a multiplication by a weighting function $\phi' = \phi'(\mathbf{x}')$, gives

$$\int_{\overline{V}}\left[\int_{dv} \phi' t'_{kl,k}\, dv' + \int_{dv} \phi'\rho'(f'_l - a'_l)\, dv\right] = 0$$

which may be written as

$$\int_{\overline{V}}\left[\int_{dv} (\phi' t'_{kl})_{,k}\, dv' + \int_{dv}\left\{t'_{kl}\phi'_{,k} + \phi'\rho'(f'_l - a'_l)\right\}dv'\right] = 0 \tag{5.5.3}$$

The first term is a volume integral which, upon using Gauss' theorem, may be converted into a surface integral such that

$$\int_{\bar{S}}\int_{ds} \phi' t'_{kl}\, da'_k = \int_{\bar{V}}\int_{dv} (\phi' t'_{kl})_{,k}\, dv'$$

where $da'_k = n'_k ds$, n'_k being the unit normal to ds at $\mathbf{x}'$. Using this result, the general form of the weighted moment of Cauchy's first law over $\bar{V}$ becomes

$$\int_{\bar{S}}\int_{ds} \phi' t'_{kl}\, da'_k + \int_{\bar{V}}\int_{dv} [- t'_{kl}\phi'_{,k} + \phi'\rho'(f'_l - a'_l)]dv' = 0 \tag{5.5.4}$$

Now consider the special case when the weighting function $\phi'(\mathbf{x}') \equiv 1$. Then, Eq. (5.5.4) can be written as

$$\int_{\bar{S}} t_{kl}\, da_k + \int_{\bar{V}} \rho(f_l - a_l)\, dv = 0 \tag{5.5.5}$$

where t_{kl} and f_l are defined by

$$\left.\begin{aligned} t_{kl}\, da_k &\equiv \int_{ds} t'_{kl}\, da'_k \\ \rho f_l\, dv &\equiv \int_{dv} \rho' f'_l\, dv' \end{aligned}\right\} \tag{5.5.6}$$

Now from Eq. (5.2.26), $\rho a_l\, dv = \int_{dv} \rho' a'_l\, dv'$. Use of Gauss' theorem, $\int_{\bar{S}} t_{kl}\, da_k = \int_{\bar{V}} t_{kl,k}\, dv$, in Eq. (5.5.5) then results in the field equation

$$t_{kl,k} + \rho(f_l - a_l) = 0 \tag{5.5.7}$$

This has the same form as Cauchy's first law, provided that t_{kl}, f_l and a_l are interpreted, respectively, as the stress tensor, the body force, and the acceleration at a macroelement. Note that this interpretation is possible only because it results in Eq. (5.5.7), in which each term has a well-established connotation for nonpolar materials. For the remainder of this chapter, t_{kl} and f_k will be regarded, respectively, as the stress tensor and the body force in the continuum at the macroelement level. Thus, the expression for the balance of linear momentum for a material with microstructure is the same as that for a nonpolar material without microstructure.

Next consider the case in which $\phi'(\mathbf{x}') = x'_m = x_m + \xi_m$. The different terms in Eq. (5.5.5) can then be reduced as follows:

$$\begin{aligned} \int_{\bar{S}}\int_{ds} \phi' t'_{kl}\, da'_k &= \int_{\bar{S}}\int_{ds} (x_m + \xi_m) t'_{kl}\, da'_k \\ &= \int_{\bar{S}} \left[x_m \int_{ds} t'_{kl}\, da'_k + \int_{ds} \xi_m t'_{kl}\, da'_k \right] \end{aligned}$$

$$= \int_{\bar{S}} x_m t_{kl}\, da_k + \int_{\bar{S}}\int_{ds} \xi_m t'_{kl}\, da'_k \tag{5.5.8}$$

where use has been made of Eq. (5.5.7) a. Also,

$$\int_{\bar{V}}\int_{dv} \left[-t'_{kl}\phi'_{,k} + \phi'\rho'(f'_l - a'_l)\right] dv'$$

$$= \int_{\bar{V}}\int_{dv} \left[-t'_{kl}\delta_{km} + \rho'(x_m + \xi_m)(f'_l - a'_l)\right] dv'$$

$$= \int_{\bar{V}}\int_{dv} \left[-t'_{ml} + x_m\rho'(f'_l - a'_l) + \rho'\xi_m(f'_l - a'_l)\right] dv'$$

$$= \int_{\bar{V}} x_m\rho(f_l - a_l)\, dv + \int_{\bar{V}}\int_{dv} \left[-t'_{ml} + \rho'\xi_m(f'_l - a'_l)\right] dv' \tag{5.5.9}$$

Substitution from Eqs. (5.5.8) and (5.5.9) in Eq. (5.5.5) then results in

$$\int_{\bar{S}} x_m t_{kl}\, da_k + \int_{\bar{S}}\int_{ds} \xi_m t'_{kl}\, da'_k + \int_{\bar{V}}\int_{dv} x_m\rho(f_l - a_l)\, dv$$
$$+ \int_{\bar{V}}\int_{dv} \left[-t'_{ml} + \rho'\xi_m(f'_l - a'_l)\right] dv = 0 \tag{5.5.10}$$

In order to interpret this equation at the macro level some macro quantities are further defined as follows:

$$\lambda_{klm}\, da_k \equiv \int_{ds} t'_{kl}\xi_m\, da'_k \tag{5.5.11}$$

$$\rho\, l_{lm}\, dv \equiv \int_{dv} \rho' f'_l \xi_m\, dv' \tag{5.5.12}$$

$$\rho\dot{\sigma}_{lm} \equiv \int_{dv} \rho' a'_l \xi_m\, dv' \tag{5.5.13}$$

$$s_{lm} \equiv \int_{dv} t'_{lm}\, dv' \tag{5.5.14}$$

With these definitions, Eq. (5.5.10) becomes

$$\int_{\bar{S}} (t_{kl}x_m + \lambda_{klm})\, da_k + \int_{\bar{V}} \left[-s_{lm} + \rho(f_l - a_l)x_m + \rho(l_{lm} - \dot{\sigma}_{lm})\right] dv = 0$$

Use of Gauss' theorem then gives

$$\int_V [x_m\{t_{kl,k} + \rho(f_l - a_l)\} + t_{ml} - s_{ml} + \lambda_{klm,k} + \rho(l_{lm} - \dot{\sigma}_{lm})]\, dv = 0$$

which results in the field equation

$$\lambda_{klm,k} + t_{ml} - s_{ml} + \rho(l_{lm} - \dot{\sigma}_{lm}) = 0 \tag{5.5.15}$$

where use has been made of Eq. (5.5.5) to eliminate the remaining terms.

In Eq. (5.5.11), the integral gives the moments of the surface tractions acting on the microsurfaces that make up the macrosurface elements ds, about its centroid $\mathbf{x}$. Hence $\lambda_{klm}\, da_k$ are the first surface stress moments. λ_{klm} will therefore be called the first stress moment. Similarly, the integrals in Eqs. (5.5.12) and (5.5.13) are, respectively, the sums of moments of body forces and inertia forces of the microelements, with respect to the center of mass $\mathbf{x}$ of the macroelement dv. Therefore, l_{lm} and $\dot{\sigma}_{lm}$ will be called the first body moment and the internal spin, respectively. The tensor s_{ml}, which by definition is symmetric, is an average of the stresses over a microvolume element and is, therefore, called the microstress average. So far, Eq. (5.5.15) has not been related to the established equations of mechanics.

The symmetric and skew-symmetric parts of Eq. (5.5.15) are given, respectively, by

$$\lambda_{k(lm),k} + t_{(ml)} - s_{ml} + \rho\left[l_{(lm)} - \dot{\sigma}_{(lm)}\right] = 0 \tag{5.5.16}$$

$$\lambda_{k[lm],k} + t_{[ml]} + \rho(l_{[lm]} - \dot{\sigma}_{[lm]}) = 0 \tag{5.5.17}$$

where parentheses and brackets enclosing pairs of indices denote, respectively, symmetric and skew-symmetric parts. For example, $\lambda_{k(lm)} = \frac{1}{2}(\lambda_{klm} + \lambda_{kml})$ and $\lambda_{k[lm]} = \frac{1}{2}(\lambda_{klm} - \lambda_{kml})$.

Identification of Eq. (5.5.17) with the equation governing the balance of angular momentum, given by Eq. (2.6.5), makes it possible to give an interpretation for λ_{klm}. Since s_{kl} is symmetric, a multiplication of Eq. (5.5.15) by e_{iml} results in

$$\rho\, e_{iml}\dot{\sigma}_{lm} = (e_{iml}\lambda_{klm})_{,k} + e_{iml}t_{ml} + \rho\, e_{iml}l_{lm}$$

This is the same as Eq. (2.6.5) provided the following identifications are made:

$$\left.\begin{aligned} \dot{\sigma}_i &= e_{irs}\dot{\sigma}_{sr} = e_{irs}\dot{\sigma}_{[sr]} \\ m_{ij} &= e_{jrs}\lambda_{isr} = e_{jrs}\lambda_{i[sr]} \\ l_i &= e_{irs}l_{sr} = e_{irs}l_{[sr]} \end{aligned}\right\} \tag{5.5.18}$$

These equations may be inverted to give

$$\left.\begin{aligned} \dot{\sigma}_{[ij]} &= \frac{1}{2} e_{jir} \dot{\sigma}_r \\ \lambda_{i[jk]} &= \frac{1}{2} e_{kjr} m_{ir} \\ l_{[ij]} &= \frac{1}{2} e_{jir} l_r \end{aligned}\right\} \tag{5.5.19}$$

The skew-symmetric part of Eq. (5.5.15), given by Eq. (5.5.17), gives the balance of angular momentum for polar materials. In addition to these equations, there are six equations, given by Eq. (5.5.16), which relate the symmetric parts $\lambda_{k(lm)}$, $t_{(ml)}$, s_{ml}, $l_{(lm)}$ and $\dot{\sigma}_{(lm)}$.

Besides the usual six equations governing the balance of linear and angular momenta for a polar material, the zeroeth and first moments of Eq. (5.5.1), given respectively by Eqs. (5.5.8) and (5.5.15), require that for a micromorphic material an additional set of six equations, given by Eq. (5.5.16), be satisfied.

Before considering the balance of energy, a discussion of the concept of microinertia is appropriate.

5.6 Microinertia Moments

Using Eqs. (5.2.14) and (5.2.22), the expression for the internal spin, given in Eq. (5.5.13), can be reduced as follows:

$$\begin{aligned} \rho\dot{\sigma}_{kl}\, dv &= \int_{dv} \rho' a'_k \xi_l \, dv = \int_{dv} \rho'(a_k + \ddot{\chi}_{Kk}\Xi_K)\chi_{Ll}\Xi_L \, dv' \\ &= a_k \chi_{Ll} \int_{dv} \rho' \Xi_L \, dv' + \ddot{\chi}_{Kk}\chi_{Ll} \int_{dv} \rho' \Xi_K \Xi_L \, dv' \\ &= a_k \chi_{Ll} \int_{dv} \mathrm{P}' \Xi_L \, dV' + \ddot{\chi}_{Kk}\chi_{Ll} \int_{dv} \mathrm{P}' \Xi_K \Xi_L \, dV' \end{aligned}$$

Since the first term on the right hand side is zero,

$$\dot{\sigma}_{kl} = I_{KL} \ddot{\chi}_{Kk} \chi_{Ll} \tag{5.6.1}$$

where

$$\mathrm{P} I_{KL} \, dV \equiv \int_{dV} \mathrm{P}' \Xi_K \Xi_L \, dV' \tag{5.6.2}$$

If

$$\rho\, i_{kl} \, dv \equiv \int_{dv} \rho' \xi_k \xi_l \, dv' \tag{5.6.3}$$

then

$$\rho\, i_{kl}\, dv = \int_{dv} \rho' \chi_{Kk}\chi_{Ll}\Xi_K\Xi_L\, dv'$$

$$= \chi_{Kk}\chi_{Ll}\int_{dV} \mathrm{P}'\Xi_K\Xi_L\, dV'$$

$$= \chi_{Kk}\chi_{Ll}\mathrm{P} I_{KL}\, dV = \chi_{Kk}\chi_{Ll} I_{KL}\rho\, dv$$

so that

$$i_{kl} = I_{KL}\chi_{Kk}\chi_{Ll} \tag{5.6.4}$$

which can be inverted to give

$$I_{KL} = i_{kl}X_{kK}X_{lL} \tag{5.6.5}$$

Equation (5.6.1) can then be written as

$$\dot{\sigma}_{kl} = I_{KL}\ddot{\chi}_{Kk}\chi_{Ll} = i_{mn}X_{mK}X_{nL}\ddot{\chi}_{Kk}\chi_{Ll}$$

$$= i_{ml}X_{mK}\ddot{\chi}_{Kk}$$

which, upon use of Eq. (5.2.24), finally gives

$$\dot{\sigma}_{kl} = i_{ml}(\dot{\nu}_{mk} + \nu_{mr}\nu_{rk}) \tag{5.6.6}$$

Physically, I_{KL} resembles the moment of inertia tensor of the macroelement with respect to its center of mass. For this reason i_{ml} is called the microinertia moment and ν_{kl} is called the gyration tensor.

An expression for the time rate of change of the microinertia tensor i_{km} is useful. A time differentiation of Eq. (5.6.4), together with the use of Eq. (5.2.23), gives

$$\frac{D}{Dt}(i_{km}) = I_{KM}\dot{\chi}_{Kk}\chi_{Mm} + I_{KM}\chi_{Kk}\dot{\chi}_{Mm}$$

$$= I_{KM}\chi_{Kl}\chi_{Mm}\nu_{lk} + I_{KM}\chi_{Kk}\chi_{Ml}\nu_{lm}$$

so that

$$\frac{D}{Dt}(i_{km}) = \frac{\partial}{\partial t}i_{km} + v_r i_{km,r} = i_{lm}\nu_{lk} + i_{kl}\nu_{lm} \tag{5.6.7}$$

5.7 Balance of Energy

The local balance of energy at a point $\mathbf{x}'$ of the deformed macroelement will be assumed to be governed by the energy equation for nonpolar materials,

$$\rho'\dot{\epsilon}' = t'_{kl}v'_{l,k} - q'_{k,k} + \rho' h' \tag{5.7.1}$$

where ϵ' and h' are, respectively the internal energy density and the heat source per unit mass of the microelement and q'_k is the energy influx rate vector at $\mathbf{x}'$.

To obtain the local balance of energy, an integration of Eq. (5.7.1) over the volume V gives

$$\int_{\overline{V}}\int_{dv} \rho'\dot{\epsilon}' \, dv' = \int_{\overline{V}}\int_{dv} t'_{kl}v'_{l,k} \, dv' - \int_{\overline{V}}\int_{dv} q'_{k,k} \, dv' + \int_{\overline{V}}\int_{dv} \rho' h' \, dv' \tag{5.7.2}$$

From the continuity equation,

$$\int_{dv} \rho'\dot{\epsilon}' \, dv' = \int_{dv} \rho' \frac{D\epsilon'}{Dt} \, dv' = \frac{D}{Dt}\int_{dv} \rho'\epsilon' \, dv' = \frac{D}{Dt}(\rho\epsilon \, dv) = \rho\dot{\epsilon} \, dv \tag{5.7.3}$$

where ϵ, the internal energy density per unit mass of the macroelement, is defined by

$$\rho\epsilon \, dv \equiv \int_{dv} \rho'\epsilon' \, dv' \tag{5.7.4}$$

Similarly, the heat source average per unit mass, h, of the macroelement is defined by

$$\rho h \, dv \equiv \int_{dv} \rho' h' \, dv' \tag{5.7.5}$$

Furthermore, by using Gauss' theorem,

$$\int_{\overline{V}}\int_{dv} q'_{k,k} \, dv' = \int_{\overline{S}}\int_{ds} q'_k \, da'_k = \int_{\overline{S}} q_k \, da_k \tag{5.7.6}$$

where the energy influx rate vector, for a macroelement, is defined by

$$q_k \, da_k \equiv \int_{ds} q'_k \, da'_k \tag{5.7.7}$$

The energy input into the microelement through its surface has been assumed to be equal to the energy output, so that the microelement is in thermal equilibrium.

The remaining terms in Eq. (5.7.2) may be reduced as follows:

$$\int_{\bar{V}}\int_{dv} t'_{kl}v'_{l,k}\, dv' = \int_{\bar{V}}\int_{dv} (t'_{kl}v'_l)_{,k}\, dv' - \int_{\bar{V}}\int_{dv} t'_{kl,k}v'_l\, dv'$$

$$= \int_{\bar{S}}\int_{ds} t'_{kl}v'_l\, da'_k - \int_{\bar{V}}\int_{dv} t'_{kl,k}v'_l\, dv' \tag{5.7.8}$$

Since $v'_l = v_l + \nu_{nl}\xi_m$,

$$\int_{\bar{S}}\int_{ds} t'_{kl}v'_l\, da'_k = \int_{\bar{S}} v_l \int_{ds} t'_{kl}\, da'_k + \int_{\bar{S}} \nu_{ml} \int_{ds} t'_{kl}\xi_m\, da'_k$$

$$= \int_{\bar{S}} (t_{kl}v_l + \lambda_{klm}\nu_{ml})\, da_k \tag{5.7.9}$$

By virtue of Eq. (5.5.1)

$$\int_{\bar{V}}\int_{dv} t'_{kl,k}v'_l\, dv' = \int_{\bar{V}}\int_{dv} [\rho'(a'_l - f'_l)(v_l + \nu_{ml}\xi_m)]\, dv'$$

$$= \int_{\bar{V}} v_l \int_{dv} \rho'(a'_l - f'_l)\, dv' + \int_{\bar{V}} \nu_{ml} \int_{dv} \rho'(a'_l - f'_l)\xi_m\, dv'$$

$$= \int_{\bar{V}} [\rho(a_l - f_l)v_l + \rho\nu_{ml}(\dot{\sigma}_{lm} - l_{lm})]\, dv \tag{5.7.10}$$

Substituting from Eqs. (5.7.9) and (5.7.10) in Eq. (5.7.8), and the resulting equation and Eqs. (5.7.4) to (5.7.7) in Eq. (5.7.2),

$$\int_{\bar{V}} \rho\dot{\epsilon}\, dv = \int_{\bar{S}} (t_{kl}v_l + \lambda_{klm}\nu_{ml} - q_k)\, da_k$$

$$+ \int_{\bar{V}} [\rho(f_l - a_l)v_l + \rho(l_{lm} - \dot{\sigma}_{lm})\nu_{ml} + \rho h]\, dv \tag{5.7.11}$$

which can be rewritten as

$$\int_{\bar{V}} [\rho\dot{\epsilon} - t_{kl}v_{l,k} - \lambda_{klm}\nu_{ml,k} - (s_{ml} - t_{ml})\nu_{ml} + q_{k,k} - \rho\, h]\, dv$$

$$- \int_{\bar{V}} [\lambda_{klm,k} + t_{ml} - s_{ml} + \rho(l_{lm} - \dot{\sigma}_{lm})]\nu_{ml}\, dv$$

$$- \int_{\bar{V}} [t_{kl,k} + \rho(f_l - a_l)]\, v_l\, dv = 0$$

Finally, since the second and third integrals are zero by virtue of Eqs. (5.5.15) and (5.5.18), respectively, the balance, or conservation, of energy is governed by the field equation

$$\rho\dot{\epsilon} = t_{kl}v_{l,k} + (s_{kl} - t_{kl})\nu_{kl} + \lambda_{klm}\nu_{ml,k} - q_{k,k} + \rho\ h \tag{5.7.12}$$

In this equation the kinematic variables $v_{l,k}$ and ν_{kl} are not objective. By using the relation $b_{kl} = \nu_{kl} + v_{kl}$,

$$\begin{aligned} t_{kl}v_{l,k} + (s_{kl} - t_{kl})\nu_{kl} &= t_{kl}(v_{l,k} - \nu_{kl}) + s_{kl}(b_{kl} - v_{k,l}) \\ &= t_{kl}[v_{l,k} + v_{k,l} - (\nu_{kl} + v_{k,l})] + s_{kl}(b_{kl} - v_{(k,l)}) \\ &= t_{kl}(2d_{kl} - b_{kl}) + s_{kl}(b_{kl} - d_{kl}) \\ &= t_{kl}d_{kl} + (s_{kl} - t_{kl})(b_{kl} - d_{kl}) \end{aligned} \tag{5.7.13}$$

where use has been made of the symmetry of s_{kl}. In terms of objective measures the energy balance equation is given by

$$\rho\dot{\epsilon} = t_{kl}d_{kl} + (s_{kl} - t_{kl})(b_{kl} - d_{kl}) + \lambda_{klm}a_{mlk} - q_{k,k} + \rho h \tag{5.7.14}$$

which can also be written as

$$\rho\dot{\epsilon} = 2t_{(kl)}d_{kl} + s_{kl}\nu_{(kl)} - t_{kl}b_{kl} + \lambda_{klm}a_{mlk} - q_{k,k} + \rho h \tag{5.7.15}$$

5.8 *Entropy Inequality*

The entropy inequality will only be discussed for a class of fluids called simple micro fluids. A simple micro fluid is assumed to possess an internal energy function ϵ that depends solely on the entropy density η, the specific volume $1/\rho$, and the microinertia i_{km}, so that

$$\epsilon = \epsilon(\eta, 1/\rho, i_{km}) \tag{5.8.1}$$

Then, from Eq. (2.9.10) the thermodynamic tensions are defined by

$$\theta = \frac{\partial\epsilon}{\partial\eta}\Big|_{\rho^{-1},\mathrm{i}}\ ,\quad \pi = -\frac{\partial\epsilon}{\partial(1/\rho)}\Big|_{\eta,\mathrm{i}}\ ,\quad \pi_{km} = \frac{\partial\epsilon}{\partial i_{km}}\Big|_{\rho^{-1},\eta} \tag{5.8.2}$$

where θ is the thermodynamic temperature, π the thermodynamic pressure, and π_{km} the thermodynamic micropressure tensor. The symmetry of π_{km} follows from the symmetry of i_{km}.

The equation $\dot{\epsilon} = \theta\dot{\eta} + \tau_\alpha\dot{\nu}_\alpha$, given by Eq. (2.9.11), can then be written as

$$\dot{\epsilon} = \theta\dot{\eta} - \pi\frac{D}{Dt}\left(\frac{1}{\rho}\right) + \pi_{km}\frac{D}{Dt}i_{km} \tag{5.8.3}$$

Now the continuity equation gives $(D/Dt)(1/\rho) = -\dot{\rho}/\rho^2 = v_{k,k}/\rho$. Also, from Eq. (5.6.7),

$$\pi_{km}\frac{D}{Dt}i_{km} = \pi_{km}i_{lm}\nu_{lk} + \pi_{km}i_{kl}\nu_{lm} = \pi_{km}i_{lm}\nu_{lk} + \pi_{mk}i_{ml}\nu_{lk} = 2\pi_{km}i_{lm}\nu_{l,k}$$

since both π_{km} and i_{ml} are symmetric. Therefore, Eq. (5.8.3) can be written as

$$\dot{\epsilon} = \theta\dot{\eta} - \frac{\pi}{\rho}v_{k,k} + 2\pi_{km}i_{lm}\nu_{lk} \tag{5.8.4}$$

A substitution for ϵ from this equation in Eq. (5.7.14) results in

$$\rho\theta\dot{\eta} = \pi v_{k,k} - 2\rho\pi_{km}i_{lm}\nu_{lk} + t_{kl}v_{l,k} + (s_{ml} - t_{ml})\nu_{ml} + \lambda_{klm}\nu_{ml,k} - q_{k,k} + \rho h \tag{5.8.5}$$

If the stresses **T** and **S** are decomposed into two parts:

$$t_{kl} = -\bar{p}\delta_{kl} + \bar{t}_{kl} \quad , \quad s_{kl} = -\bar{p}\delta_{kl} + \bar{s}_{kl} \tag{5.8.6}$$

where $\bar{p}$ represents a hydrostatic pressure, then Eq. (5.8.5) can be rewritten as

$$\rho\theta\dot{\eta} = (\pi - \bar{p})v_{k,k} + \bar{t}_{kl}v_{l,k} - (\bar{s}_{kl} - \bar{t}_{kl} - \tau_{kl})\nu_{kl} + \lambda_{klm}\nu_{ml,k} - q_{k,k} + \rho h \tag{5.8.7}$$

where $\tau_{kl} = 2\pi i_{km}\tau_{lm}$ may be called the thermodynamic microstress tensor.

From Eq. (5.8.7), the dissipative "forces," which contribute to the time rate of change of entropy, are (i) the difference between the mechanical and thermodynamic pressures, (ii) the stress power, (iii) the difference of microstress and thermodynamic microstress from the stress, (iv) the stress moments, and (v) the heat influx and heat source.

Note that the micro fluid with no rigid structure possesses a reversible thermodynamic stress whose energy must be subtracted from the stress energies in calculating the entropy production. This reversible energy is not encountered in the classical Stokesian fluids.

Use of Eqs. (2.9.3) and (5.8.7) gives

$$\rho\theta\gamma = (\pi - \bar{p})v_{k,k} + \bar{t}_{kl}v_{l,k} + (\bar{s}_{kl} - \bar{t}_{kl} - \tau_{kl})\nu_{kl} + \lambda_{klm}\nu_{ml,k} - l_k(\ln\theta)_{,k} \tag{5.8.8}$$

The Clausius-Duhem inequality then requires that

$$\gamma \geqslant 0 \tag{5.8.9}$$

5.9 Constitutive Equations for Micro Fluids

A fluent medium will be called a simple micro fluid if it possesses constitutive equations of the form

$$\left.\begin{aligned} t_{kl} &= f_{kl}(v_{p,q}, \nu_{pq}, \nu_{pq,r}) \\ s_{kl} &= g_{kl}(v_{p,q}, \nu_{pq}, \nu_{pr,q}) \\ \lambda_{klm} &= h_{klm}(v_{p,q}, \nu_{pq}, \nu_{pq,r}) \end{aligned}\right\} \tag{5.9.1}$$

subject to spatial and material objectivity, and if

$$t_{kl} = s_{kl} = -\pi\delta_{kl} \quad when \quad b_{kl} = b_{kl} = d_{kl} = 0 \tag{5.9.2}$$

that is, when there is no motion other than a uniform one.

According to the principle of objectivity, Eq. (5.9.1) must be form invariant in any two objectively equivalent motions $\hat{\mathbf{x}}'$ and $\mathbf{x}'$, related to each other by $\hat{x}'_k = Q_{kl}x'_l + b'_k$, $Q_{kr}Q_{lr} = \delta_{kl}$, where Q_{kl} and b'_k are functions of time only. This imposes conditions on the forms of the functions f_{kl}, g_{kl} and h_{klm}. In order to apply the principle of objectivity, consider two rectangular frames $\hat{\mathbf{x}}'$ and $\mathbf{x}'$ connected by Eq. (5.3.9), or equivalently by Eq. (5.3.11), so that

$$\hat{x}'_k = Q_{kl}x'_l + b'_k \quad , \quad \hat{\chi}_{Kk} = Q_{kl}\chi_{Kl} \tag{5.9.3}$$

Then, form invariance of f_{kl} requires that, in the frame $\hat{\mathbf{x}}'$,

$$\hat{t}_{kl} = f_{kl}(\hat{v}_{p,q}, \hat{\nu}_{pq}, \hat{a}_{pq,r}) \tag{5.9.4}$$

Similar expressions hold for $\hat{s}_{kl}$ and $\hat{\lambda}_{klm}$. Also, from Eqs. (5.3.10), (5.3.13) and (5.3.14), and the objective nature of the stress,

$$\left.\begin{aligned} \hat{t}_{kl} &= Q_{km}Q_{ln}t_{mn} \\ \hat{v}_{k,l} &= Q_{km}Q_{ln}v_{m,n} + \dot{Q}_{kr}Q_{lr} \\ \hat{\nu}_{kl} &= Q_{km}Q_{ln}\nu_{mn} + Q_{kr}\dot{Q}_{lr} \\ \hat{\nu}_{kl,m} &= Q_{kr}Q_{ls}Q_{mt}\nu_{rs,t} \end{aligned}\right\} \tag{5.9.5}$$

Thus, from Eqs. (5.9.1), (5.9.4) and (5.9.5),

$$\begin{aligned} &\mathbf{QF}(v_{p,q}, \nu_{p,q}, \nu_{pq,r})\mathbf{Q}^T \\ &\qquad = \mathbf{F}(Q_{km}Q_{ln}v_{m,n} + \dot{Q}_{kr}Q_{lr} \ , \ Q_{km}Q_{ln}\nu_{mn} + Q_{kr}\dot{Q}_{lr} \ , \\ &\qquad\qquad Q_{kr}Q_{ls}Q_{mt}\nu_{rs,t}) \end{aligned} \tag{5.9.6}$$

must be valid for all orthogonal tensors **Q**.

Since **Q** is orthogonal, $\mathbf{Q}\dot{\mathbf{Q}}^T = (Q_{kr}\dot{Q}_{lr})$ is skew-symmetric. Therefore **Q** can always be chosen such that $\mathbf{Q} = \mathbf{I}$ and $\dot{\mathbf{Q}}$ is equal to an arbitrary skew-symmetric tensor. Upon choosing $Q_{kl} = \delta_{kl}$ and $\dot{Q}_{kr} = \frac{1}{2}(v_{r,k} - v_{k,r})$, Eq. (5.9.6) reduces to

$$\mathbf{F}(v_{p,q}, \nu_{pq}, \nu_{pq,r}) = \mathbf{F}(d_{kl}, b_{kl} - d_{kl}, a_{klm}) \tag{5.9.7}$$

Similar results can be shown to be true for **S** and **Λ**. Thus it follows that

$$\left.\begin{aligned} \mathbf{T} &= \mathbf{F}(\mathbf{D}, \mathbf{B} - \mathbf{D}, \mathbf{A}) \\ \mathbf{S} &= \mathbf{G}(\mathbf{D}, \mathbf{B} - \mathbf{D}, \mathbf{A}) \\ \mathbf{\Lambda} &= \mathbf{H}(\mathbf{D}, \mathbf{B} - \mathbf{D}, \mathbf{A}) \end{aligned}\right\} \tag{5.9.8}$$

The arguments **D**, **B** − **D** and **A** of **F**, **G** and **H** are now objective tensors. Since **B** is objective, **D**, **B** and **A** could be used as the arguments. However, the choice **D**, **B** − **D**, **A** is more useful when the functions are to be expanded around **B** − **D**.

In order to satisfy the requirements of objectivity, indicated in Eq. (5.9.6), Eqs. (5.9.8) must, for all orthogonal **Q**, satisfy

$$\left.\begin{aligned} &\mathbf{F}(\mathbf{QDQ}^T, Q(\mathbf{B} - \mathbf{D})\mathbf{Q}^T, \mathbf{QAQ}^T\mathbf{Q}^T) = \mathbf{QP}(\mathbf{D}, \mathbf{B} - \mathbf{D}, \mathbf{A})\mathbf{Q}^T \\ &\mathbf{G}(\mathbf{QDQ}^T, \mathbf{Q}(\mathbf{B} - \mathbf{D})\mathbf{Q}^T, \mathbf{QAQ}^T\mathbf{Q}^T) = \mathbf{QG}(\mathbf{D}, \mathbf{B} - \mathbf{D}, \mathbf{A})Q^T \\ &\mathbf{H}(\mathbf{QDQ}^T, \mathbf{Q}(\mathbf{B} - \mathbf{D})\mathbf{Q}^T, \mathbf{QAQ}^T\mathbf{Q}^T) = \mathbf{QH}(\mathbf{D}, \mathbf{B} - \mathbf{D}, \mathbf{A})\mathbf{Q}^T\mathbf{Q}^T \end{aligned}\right\} \tag{5.9.9}$$

where $\mathbf{QAQ}^T\mathbf{Q}^T$ stands for $Q_{kr}Q_{ls}Q_{mt}a_{rst}$. A choice of $\mathbf{Q} = -\mathbf{I}$ gives

$$\left.\begin{aligned} \mathbf{F}(\mathbf{D}, \mathbf{B} - \mathbf{D}, -\mathbf{A}) &= \mathbf{F}(\mathbf{D}, \mathbf{B} - \mathbf{D}, \mathbf{A}) \\ \mathbf{G}(\mathbf{D}, \mathbf{B} - \mathbf{D}, -\mathbf{A}) &= \mathbf{G}(\mathbf{D}, \mathbf{B} - \mathbf{D}, \mathbf{A}) \\ \mathbf{H}(\mathbf{D}, \mathbf{B} - \mathbf{D}, -\mathbf{A}) &= -\mathbf{H}(\mathbf{D}, \mathbf{B} - \mathbf{D}, \mathbf{A}) \end{aligned}\right\} \tag{5.9.10}$$

thereby showing that **F** and **G** are even in **A** while **H** is odd in **A**. Thus

$$\mathbf{H}(\mathbf{D}, \mathbf{B} - \mathbf{D}, \mathbf{0}) = 0 \tag{5.9.11}$$

so that the stress moments, $\mathbf{\Lambda} = (\lambda_{klm})$, must vanish with the vanishing of the microdeformation rate tensor **A**.

The constitutive equations for simple micro fluids may then be written as

$$\left.\begin{aligned} t_{kl} &= f_{kl}(d_{pq}, b_{pq} - d_{pq}, a_{pqr}) \\ s_{kl} &= q_{kl}(d_{pq}, b_{pq} - d_{pq}, a_{pqr}) \\ \lambda_{klm} &= h_{klm}(d_{pq}, b_{pq} - d_{pq}, a_{pqr}) \end{aligned}\right\} \tag{5.9.12}$$

where **F**, **G** and **H** are subject to Eqs. (5.9.9), (5.9.11), and

$$\left.\begin{aligned} f_{kl}(0,0,0) &= -\pi\delta_{kl} \\ g_{kl}(0,0,0) &= -\pi\delta_{kl} \\ h_{klm}(0,0,0) &= 0 \end{aligned}\right\} \tag{5.9.13}$$

The conditions on the forms of **F**, **G** and **H**, given in Eqs. (5.9.9), are similar to conditions required of isotropic tensors. Since third order tensors are involved, the determination of the complete invariants of **D**, **B** and **A** is not easy. However, since the microrotations represented by the tensors **B** − **D** and **A** are generally small, a power series representation of the constitutive equations in **B** − **D** and **A**, stopping at the first order terms, may be resorted to. Thus

$$\left.\begin{aligned} t_{lk} &= f^0_{kl}(\mathbf{D}, \mathbf{B}-\mathbf{D}, \mathbf{B}^T-\mathbf{D}) + 0(\mathbf{A}^2) \\ s_{kl} &= g^0_{kl}(\mathbf{D}, \mathbf{B}-\mathbf{D}, \mathbf{B}^T-\mathbf{D}) + 0(\mathbf{A}^2) \\ \lambda_{klm} &= h^0_{klm}(\mathbf{D}, \mathbf{B}-\mathbf{D}, \mathbf{B}^T-\mathbf{D}) \\ &\quad + h^1_{klmrst}(\mathbf{D}, \mathbf{B}-\mathbf{D}, \mathbf{B}^T-\mathbf{D})\,a_{rst} + 0(\mathbf{A}^3) \end{aligned}\right\} \tag{5.9.14}$$

By virtue of Eqs. (5.9.10) a, b, the expressions for **T** and **S** do not have linear terms in **A**. The inclusion of $\mathbf{B}^T - \mathbf{D}$, the transpose of **B** − **D**, into the arguments of these functions is required for the purpose of making these functions isotropic, since **T**, **B** and **A** are not, in general, symmetric tensors. If the tensor functions $\mathbf{F}^0$, $\mathbf{G}^0$, $\mathbf{H}^0$ and $\mathbf{H}^1$ are further assumed to be polynomials in the matrices **D**, **B** − **D** and $\mathbf{B}^T - \mathbf{D}$, then they can be expressed in a finite number of terms. Using results due to Spencer and Rivlin [Ref. 3], **F** and **G** are expressible as polynomials, each having 85 terms. Here only three terms, which contain polynomials that are linear in **B** − **D** and $\mathbf{B}^T - \mathbf{D}$, will be considered. Then

$$\begin{aligned} \mathbf{T} = {} & \alpha_0\mathbf{I} + \alpha_1\mathbf{D} + \alpha_2\mathbf{D}^2 + \alpha_3(\mathbf{B}-\mathbf{D}) + \alpha_4(\mathbf{B}^T-\mathbf{D}) \\ & + \alpha_5\mathbf{D}(\mathbf{B}-\mathbf{D}) + \alpha_6(\mathbf{B}-\mathbf{D})\mathbf{D} + \alpha_7\mathbf{D}(\mathbf{B}^T-\mathbf{D}) + \alpha_8(\mathbf{B}^T-\mathbf{D})\mathbf{D} \\ & + \alpha_9\mathbf{D}^2(\mathbf{B}-\mathbf{D}) + \alpha_{10}(\mathbf{B}-\mathbf{D})\mathbf{D}^2 \\ & + \alpha_{11}\mathbf{D}^2(\mathbf{B}^T-\mathbf{D}) + \alpha_{12}(\mathbf{B}^T-\mathbf{D})\mathbf{D}^2 \\ & + \alpha_{13}\mathbf{D}(\mathbf{B}-\mathbf{D})\mathbf{D}^2 + \alpha_{14}\mathbf{D}(\mathbf{B}^T-\mathbf{D})\mathbf{D}^2 \end{aligned} \tag{5.9.15}$$

An identical expression with α_k replaced by β'_k, $k = 0, 1, \cdots, 14$, is valid for **S**. The coefficients α_k, β'_k, for $k = 0, 1, 2$, are polynomials of the six invariants:

$$\begin{array}{l} \mathrm{tr}\,\mathbf{D}\ ,\ \mathrm{tr}\,\mathbf{D}^2\ ,\ \mathrm{tr}\,\mathbf{D}^3 \\ \mathrm{tr}(\mathbf{B}-\mathbf{D})\ ,\ \mathrm{tr}(\mathbf{B}-\mathbf{D})\mathbf{D}\ ,\ \mathrm{tr}(\mathbf{B}-\mathbf{D})\mathbf{D}^2 \end{array} \tag{5.9.16}$$

being linear in the last three, and α_k and β'_k $k = 4, 5, \cdots, 14$, are functions of the first three invariants alone. That is

$$\left.\begin{array}{l} \alpha_k = \alpha_{k0} + \alpha_{k1}\mathrm{tr}(\mathbf{B}-\mathbf{D}) + \alpha_{k2}\mathrm{tr}(\mathbf{B}-\mathbf{D})\mathbf{D} \\ \qquad\qquad + \alpha_{k3}\mathrm{tr}(\mathbf{B}-\mathbf{D})\mathbf{D}^2 \\ \alpha_{k\lambda} = \alpha_{k\lambda}(\mathrm{tr}\,\mathbf{D}, \mathrm{tr}\,\mathbf{D}^2, \mathrm{tr}\,\mathbf{D}^3), \\ \qquad k = 0, 1, 2,\ \ \lambda = 0, 1, 2, 3 \end{array}\right\} \tag{5.9.17}$$

The symmetry of $\mathbf{S}$ can be used for further reducing the expressions for it to

$$\begin{aligned} \mathbf{S} = {} & \beta_0\mathbf{I} + \beta_1\mathbf{D} + \beta_2\,\mathbf{D}^2 + \beta_3(\mathbf{B} + \mathbf{B}^T - 2\mathbf{D}) \\ & + \beta_4(\mathbf{D}\mathbf{B} + \mathbf{B}^T\mathbf{D} - 2\mathbf{D}^2) + \beta_5(\mathbf{B}\mathbf{D} + \mathbf{D}\mathbf{B}^T - 2\mathbf{D}^2) \\ & + \beta_6(\mathbf{D}^2\mathbf{B} + \mathbf{B}^T\mathbf{D}^2 - 2\mathbf{D}^3) + \beta_7(\mathbf{B}\mathbf{D}^2 + \mathbf{D}^2\mathbf{B} - 2\mathbf{D}^3) \\ & + \beta_8(\mathbf{D}\mathbf{B}\mathbf{D}^2 + \mathbf{D}^2\mathbf{B}^T\mathbf{D} - 2\mathbf{D}^4) + \beta_9(\mathbf{D}\mathbf{B}^T\mathbf{D}^2 + \mathbf{D}^2\mathbf{B}\mathbf{D} - 2\mathbf{D}^4) \end{aligned} \tag{5.9.18}$$

where β_0, β_1 and β_2 have the same functional form as in Eq. (5.9.17) with the coefficients, $\alpha_{k\lambda}$ replaced by $\beta_{k\lambda}$, and $\beta_4, \cdots, \beta_9$ are polynomials in the first three invariants given in Eq. (5.9.16).

The condition $\mathbf{H}(\mathbf{D}, \mathbf{B}-\mathbf{D}, \mathbf{0}) = 0$, given in Eq. (5.9.10), when applied to Eq. (5.9.14) c, gives

$$h^0_{klm} = 0 \tag{5.9.19}$$

Furthermore, h^1_{klmrst} is an isotropic tensor of order six, so that upon substituting the known expression for an isotropic tensor of order six, Eq. (5.9.14) c can be written as

$$\begin{aligned} \lambda_{klm} = {} & (\gamma_1 a_{mrr} + \gamma_2 a_{rmr} + \gamma_3 a_{rrm})\delta_{kl} \\ & + (\gamma_4 a_{lrr} + \gamma_5 a_{rlr} + \gamma_6 a_{rrl})\delta_{km} \\ & + (\gamma_7 a_{krr} + \gamma_8 a_{rkr} + \gamma_9 a_{rrk})\delta_{lm} \\ & + \gamma_{10} a_{klm} + \gamma_{11} a_{kml} + \gamma_{12} a_{lkm} \\ & + \gamma_{13} a_{mkl} + \gamma_{14} a_{lmk} + \gamma_{15} a_{mkl} \end{aligned} \tag{5.9.20}$$

where

$$\gamma_k = \gamma_k(\mathbf{D}, \mathbf{B} - \mathbf{D}, \mathbf{B}^T - \mathbf{D}) \quad , \quad k = 1, 2, \cdots, 15 \tag{5.9.21}$$

The γ_k's are also scalar invariants under all orthogonal transformations $\mathbf{Q}$. Therefore, for the microlinear case under consideration, they are expressible as polynomials in the first three of the six invariants listed in Eq. (5.9.16), provided they are analytic in their arguments, so that

$$\gamma_k = \gamma_k(\mathrm{tr}\,\mathbf{D}, \mathrm{tr}\,\mathbf{D}^2, \mathrm{tr}\,\mathbf{D}^3) \quad , \quad k = 1, 2, \cdots, 15 \tag{5.9.22}$$

The conditions given by Eq. (5.9.13), namely, $f_{kl}(0,0,0) = -\pi\delta_{kl}$, $g_{lk}(0,0,0) = -\pi\delta_{kl}$ and $h_{klm}(0,0,0) = 0$, are satisfied by taking

$$\left.\begin{array}{l} \alpha_0 = -\pi + \alpha(\mathbf{D}, \mathbf{B} - \mathbf{D}, \mathbf{B}^T - \mathbf{D}) \\ \beta_0 = -\pi + \beta(\mathbf{D}, \mathbf{B} - \mathbf{D}, \mathbf{B}^T - \mathbf{D}) \\ \gamma_k(0,0,0) = \alpha(0,0,0) = \beta(0,0,0) = 0 \, ; \quad k = 1, 2, \cdots, 15 \end{array}\right\} \tag{5.9.23}$$

where α and β are functions of the six invariants listed in Eq. (5.9.16).

Before discussing fully linearized constitutive equations, all Stokesian fluids will be shown to be included in the class of simple micro fluids represented by the constitutive equations given in Eqs. (5.9.15), (5.9.17) and (5.9.18) subject to Eq. (5.9.23). To show this, take all $\alpha_k = \beta_k = 0$ for $k \geqslant 3$ and $\gamma_k = 0$ for all k. Then Eqs. (5.9.15), (5.9.17) and (5.9.18) reduce to

$$\left.\begin{array}{l} \mathbf{T} = (-\pi + \alpha)\mathbf{I} + \alpha_1\mathbf{D} + \alpha_2\mathbf{D}^2 \\ \mathbf{S} = (-\pi + \alpha)\mathbf{I} + \beta_1\mathbf{D} + \beta_2\mathbf{D}^2 \end{array}\right\} \tag{5.9.24}$$

where α, α_1, α_2, β, β_1 and β_2 are now considered to be functions of the three invariants

$$\mathrm{tr}\,\mathbf{D} \; , \; \mathrm{tr}\,\mathbf{D}^2 \; , \; \mathrm{tr}\,\mathbf{D}^3 \tag{5.9.25}$$

or equivalently

$$I_\mathbf{D} = \mathrm{tr}\mathbf{D} \; , \; II_\mathbf{D} = \frac{1}{2}(\mathrm{tr}^2\mathbf{D} - \mathrm{tr}\mathbf{D}^2) \; , \; III_\mathbf{D} = \det\mathbf{D} \tag{5.9.26}$$

If the coefficients are now chosen such that $\alpha = \beta$, $\alpha_1 = \beta_1$ and $\alpha_2 = \beta_2$, then $\mathbf{T} = \mathbf{S}$. Furthermore when l_{rs} and i_{rs} are taken to be zero, then $\dot{\boldsymbol{\sigma}} = 0$ and all the balance equations, given by Eqs. (5.5.15) and (5.6.7) are automatically satisfied, and Eqs. (5.4.1), (5.5.8) and (5.7.14) reduce to those for Stokesian fluids.

For special types of body and surface moments, and for $\mathbf{D} = \mathbf{B}$, all motions of simple micro fluids can be shown to coincide with those of Stokesian fluids. For, with $\mathbf{D} = \mathbf{B}$, the constitutive equations for the stresses reduce to Eqs. (5.9.24) a, b. Now $\mathbf{D} = \mathbf{B}$ implies that $\nu_{kl} = w_{kl} = d_{kl} - v_{k,l}$. The stress moments, $\mathbf{\Lambda}$, are then fully determined from Eq. (5.9.20) by putting

$$a_{klm} = w_{kl,m} = d_{kl,m} - v_{k,lm} \tag{5.9.27}$$

Thus the balance equations in Eqs. (5.5.15) give a special distribution for l_{rs}, and the boundary conditions $\lambda_{klm} n_k = \lambda_{lm}(\mathbf{n})$ give a special surface moment distribution $\lambda_{lm}(\mathbf{n})$. In this case, with $\alpha = \beta$, $\alpha_1 = \beta_1$ and $\alpha_2 = \beta_2$, the remaining equations are the basic equations and boundary conditions of the Stokesian theory.

5.10 Linear Theory of Micro Fluids

An expansion of the constitutive coefficients α_k, β_k and γ_k into power series of their arguments, together with a retention of only the linear terms in $\mathbf{D}$ and $\mathbf{B}$, results in

$$\left.\begin{aligned} \mathbf{T} &= [-\pi + \lambda_v \operatorname{tr} \mathbf{D} + \lambda_0 \operatorname{tr}(\mathbf{B} - \mathbf{D})]\mathbf{I} + 2\mu_v \mathbf{D} \\ &\qquad + 2\mu_0(\mathbf{B} - \mathbf{D}) + 2\mu_1(\mathbf{B}^T - \mathbf{D}) \\ \mathbf{S} &= [-\pi + \eta_v \operatorname{tr} \mathbf{D} + \eta_0 \operatorname{tr}(\mathbf{B} - \mathbf{D})]\mathbf{I} + \zeta_v \mathbf{D} + \zeta_1(\mathbf{B} + \mathbf{B}^T - 2\mathbf{D}) \end{aligned}\right\} \tag{5.10.1}$$

where λ_v, λ_0, μ_v, μ_0, η_v, η_0, ζ_v and ζ_1 are viscosity coefficients. They are, in general, functions of temperature. In order to have Stokesian fluids included in the linear theory,

$$\lambda_v = \eta_v \quad , \quad \mu_v = \zeta_v \tag{5.10.2}$$

Thus, the linear theory of simple micro fluids introduces five additional viscosities into the constitutive equations for stress. The form of the constitutive equations for stress moments is identical with Eq. (5.9.20) except that the γ_k's are now constants, or, in general, functions of the temperature alone. Including the gyroviscosities γ_k, the total number of viscosity coefficients is 22.

5.11 Equations of Motion

The equations of motion are now, the continuity equation

$$\dot{\rho} + \rho v_{k,k} = 0 \tag{5.11.1}$$

the conservation law for microinertia

$$\frac{D}{Dt}(i_{km}) = i_{rm}\nu_{rk} + i_{kr}\nu_{rm} \tag{5.11.2}$$

and the equations obtained by substituting the linear constitutive equations in the two equations of motion.

From Eq. (5.10.1)

$$t_{kl} = [-\pi + \lambda_v v_{r,r} + \lambda_0 \nu_{rr}]\delta_{kl} + \mu_v(v_{k,l} + v_{l,k})$$
$$+ \mu_0(2\nu_{kl} + v_{k,l} - v_{l,k}) + \mu_1(2\nu_{kl} + v_{l,k} - v_{k,l})$$

so that

$$t_{kl,k} = -\pi_{,l} + (\lambda_v + \mu_v + \mu_0 - \mu_1)v_{k,lk} + (\mu_v - \mu_0 + \mu_1)v_{l,kk}$$
$$+ \lambda_0\nu_{kk,l} + 2\mu_0\nu_{kl,k} + 2\mu_1\nu_{lk,k} \tag{5.11.3}$$

Thus, Eq. (5.5.8) becomes

$$-\pi_{,l} + (\lambda_v + \mu_v + \mu_0 - \mu_1)v_{k,lk} + (\mu_v - \mu_0 + \mu_1)v_{l,kk}$$
$$+ \lambda_0\nu_{kk,l} + 2\mu_0\nu_{kl,k} + 2\mu_1\nu_{lk,k} + \rho(f_l - \dot{v}_l) = 0 \tag{5.11.4}$$

or, in Gibbsian notation,

$$\rho\mathbf{a} = -\nabla\pi + (\lambda_v + \mu_v + \mu_0 - \mu_1)\nabla\nabla\cdot\mathbf{v} + (\mu_v - \mu_0 + \mu_1)\nabla^2\mathbf{v}$$
$$+ \lambda_1\nabla(\mathrm{tr}\,\mathbf{N}) + 2\mu_0\,\nabla\cdot\mathbf{N} + 2\mu_1(\mathbf{N}\cdot\nabla) + \rho\mathbf{f}$$

where $\mathbf{N} = (\nu_{kl})$.

Similarly, upon using the constitutive equations given in Eqs. (5.10.1) and (5.9.20), Eq. (5.5.15) becomes

$$(\mu_1 - \mu_1)(v_{k,l} - v_{l,k}) + (\lambda_0 - \eta_0)\nu_{rr}\,\delta_{kl} + (2\mu_0 - \zeta_1)\nu_{kl}$$
$$+ (2\mu_1 - \zeta_1)\nu_{lk} + (\gamma_1 + \gamma_{13})\nu_{km,ml} + (\gamma_2 + \gamma_{11})\nu_{mk,ml}$$
$$+ (\gamma_3 + \gamma_6)\nu_{mm,lk} + (\gamma_4 + \gamma_{12})\nu_{lm,mk} + (\gamma_5 + \gamma_{10})\nu_{ml,km} \tag{5.11.5}$$
$$+ \gamma_{14}\,\nu_{lk,mm} + \gamma_{15}\,\nu_{kl,mm}$$
$$+ (\gamma_7\,\nu_{mn,nm} + \gamma_8\,\nu_{nm,nm} + \gamma_9\,\nu_{mm,nn})\delta_{kl} + \rho(l_{lk} - \dot{\sigma}_{lk}) = 0$$

or, in Gibbsian notation

$$\rho\dot{\boldsymbol{\Sigma}} = (\mu_0 - \mu_1)(\nabla\mathbf{v} - \mathbf{v}\nabla) + (\lambda_0 - \eta_0)(\text{tr}\,\mathbf{N})\mathbf{I} + (2\mu_0 - \zeta_1)\mathbf{N}^T$$

$$+ (2\mu_1 - \zeta_1)\mathbf{N} + (\gamma_1 + \gamma_{13})\nabla(\mathbf{N}\cdot\nabla) + (\gamma_2 + \gamma_{11})\nabla(\nabla\cdot\mathbf{N})$$

$$+ (\gamma_3 + \gamma_6)\nabla\nabla(\text{tr}\mathbf{N}) + (\gamma_4 + \gamma_{12})(\mathbf{N}\cdot\nabla)\nabla + (\gamma_5 + \gamma_{10})(\nabla\cdot\mathbf{N})\nabla$$

$$+ \gamma_{14}\,\nabla^2\mathbf{N} + \gamma_{15}\,\nabla^2\mathbf{N}^T$$

$$+ [(\gamma_7 + \gamma_8)\nabla\cdot\mathbf{N}\cdot\nabla + \gamma_9\,\nabla^2(\text{tr}\mathbf{N})]\mathbf{I} + \rho\mathbf{L}$$

where $\boldsymbol{\Sigma} = (\sigma_{kl})$ and $\mathbf{L} = (l_{kl})$.

There are now 1 (continuity) + 6 (balance of microinertia) + 3 (balance of linear momentum) + 9 (balance of first stress moments) = 19 equations for determining the 19 unknowns: ρ, $i_{km} = i_{mk}$, ν_{kl} and v_k, since the body force f_l and the first body moments l_{lm} are supposed to be given and σ_{kl} is expressible in terms of i_{km} and ν_{kl}, through Eq. (5.6.6), as

$$\dot{\sigma}_{kl} = i_{ml}(\dot{\nu}_{mk} + \nu_{mr}\nu_{rk}) \tag{5.11.6}$$

Under appropriate boundary conditions, such as $t_{kl}n_k = t_l(\mathbf{n})$, $\lambda_{klm}n_k = \lambda_{lm}(\mathbf{n})$ on the boundary, and initial conditions, the complete behavior of the constitutively linear theory of micro fluids should be derivable from the foregoing partial differential equations. As initial conditions, the values of ρ, i_{km}, ν_{kl} and v_k may be prescribed at time $t = 0$. Finally, if the viscosities are chosen such that $\lambda_0 = \mu_0 = \mu_1 = \eta_0 = \zeta_1 = i_{km} = \gamma_k = 0$, then Eq. (5.11.4) reduces to the Navier-Stokes equations, the continuity equation remains valid, and Eqs. (5.11.2) and (5.11.5) reduce to identities $0 = 0$.

The linear theory of micro fluids is almost untractable. Therefore, the next step is to make further simplifications by placing restrictions on the micromotions. This can be achieved by assuming special forms for the gyration tensor.

5.12 Concluding Remarks

This chapter has examined Eringen's theory of micromorphic fluids. Each macrovolume element is assumed to contain substructures at the micro level, where the equations of the nonpolar theory are assumed to apply. The equations, which apply at the macro level, are then obtained by a suitable averaging procedure. In this way, both microstructure and couple stresses can be accounted for in a systematic way.

In addition to the usual kinematic variables, a description of the kinematics of a fluid with microstructure introduces a new tensor, namely the gyration tensor ν_{ij}.

In a fluid with fine suspended particles, the particles could be looked upon as a microstructure in the main fluid. Similarly, in a turbulent flow the eddies, or vortices, may be looked upon as a microstructure in the main fluid. Thus, such theories may be expected to have a wide range of applications.

Even the linear theory of micro fluids, discussed in Sections (5.10) and (5.11), is almost untractable. Simpler theories, such as the theory of micropolar fluids, which are more tractable, can be obtained by restricting the possible forms of the gyration tensor. Physically, this amounts to placing restrictions on the allowable forms that the microstructure can have. One such simple theory is discussed in the next chapter.

5.13 References

The material in this chapter is based on the first two references given below.

1. Eringen, A. C., and Suhubi, E. S. (1964) Nonlinear Theory of Simple Micro-Elastic Solids — I, *Int. J. Eng. Sci.* **2**, 189-203.
2. Eringen, A. C. (1964). Simple Micro-Fluids, *Int. J. Eng. Sci.* **2**, 205-217.
3. Spencer, A. J. M., and Rivlin, R. S. (1958). The Theory of Matrix Polynomials and Its Application to the Mechanics of Isotropic Continua, *Arch. Ration. Mech. Anal.* **2**, 309-435.

CHAPTER 6

Micropolar Fluids

6.1 Introduction

The theory of micro fluids, which has been presented in Chapter 5, is very general and allows for a wide variety of microstructures through the gyration tensor ν_{ij}. The simplest subclass of micro fluids in which microstructure is still present, and which is obtained by restricting the form of the gyration tensor, is the class of micropolar fluids of A. C. Eringen.

A fluid will be called micropolar if, for all motions,

$$\lambda_{klm} = -\lambda_{kml} \quad , \quad \nu_{kl} = -\nu_{lk} \quad . \tag{6.1.1}$$

Such fluids exhibit only microrotational effects and can support surface and body couples.

6.2 Skew-Symmetry of the Gyration Tensor and Microisotropy

In the general case of a micro fluid, the gyration tensor has nine independent components. For micropolar fluids the assumption of the skew-symmetry of ν_{ij} reduces the number of independent components to three, so that, in addition to their usual convection due to the motion of the fluid element, points contained in a small element of fluid can rotate about the centroid of the volume element in the average sense described by the gyration tensor $\nu_{ij} = -\nu_{ji}$. Equation (5.3.7) shows that microstretch of particles is not possible. For, if only motion about the centroid is to be considered then, by taking dx_k to be zero,

$$\frac{D}{Dt}(ds')^2 = (\nu_{kl} + \nu_{lk})\, d\xi_k d\xi_1 = 2\nu_{(kl)}\, d\xi_k d\xi_l \tag{6.2.1}$$

so that $(D/Dt)(ds')^2 = 0$ since $\nu_{(kl)} = 0$. Thus $(ds')^2$ is a constant, thereby implying that the micromotion consists purely of rotations about the centroid of the mass element.

The skew-symmetry of ν_{ij} also implies that

$$a_{klm} = -a_{kml} \tag{6.2.2}$$

A micro fluid is said to be microisotropic if the microinertia moment is isotropic, that is if

$$i_{km} = i\delta_{km} \tag{6.2.3}$$

Then, from Eq. (5.6.3), for a microisotropic fluid

$$\frac{D}{Dt}(i_{km}) = i(\delta_{rm}\nu_{rk} + \delta_{kr}\nu_{rm}) = 2i\nu_{(km)} = 0$$

when $\nu_{km} = -\nu_{mk}$, so that

$$\frac{D}{Dt}(i_{km}) = 0 \quad or \quad \frac{D}{Dt}(i) = 0 \tag{6.2.4}$$

Thus, on material lines,

$$i = \frac{1}{2}j = constant \tag{6.2.5}$$

Then, from Eq. (5.11.6),

$$\dot{\sigma}_{kl} = i(\dot{\nu}_{lk} + \nu_{lr}\nu_{rk}) \tag{6.2.6}$$

and, since $\dot{\nu}_{kl}$ and $\nu_{lr}\ \nu_{rk}$ are skew-symmetric and symmetric, respectively,

$$\dot{\sigma}_{(kl)} = i\nu_{kr}\nu_{rl} \tag{6.2.7}$$

$$\dot{\sigma}_{[kl]} = i\dot{\nu}_{lk} \tag{6.2.8}$$

6.3 Micropolar Fluids

The purpose of this section is to show that micropolar fluids are a subclass of micro fluids. That is, to show that a class of fluids satisfying

$$\lambda_{klm} = -\lambda_{kml} \quad , \quad \nu_{kl} = -\nu_{lk} \tag{6.3.1}$$

exists as a subclass of micro fluids.

From Eq. (5.9.20)

$$\begin{aligned}\lambda_{klm} + \lambda_{kml} = {} & [(\gamma_1 + \gamma_4)a_{mrr} + (\gamma_2 + \gamma_5)a_{rmr} + (\gamma_3 + \gamma_6)a_{rrm}]\delta_{kl} \\ & + [(\gamma_1 + \gamma_4)a_{lrr} + (\gamma_2 + \gamma_5)a_{rlr} + (\gamma_3 + \gamma_6)a_{rrl}]\delta_{km} \\ & + 2[\gamma_7 a_{krr} + \gamma_8 a_{rkr} + \gamma_9\ a_{rrk}]\delta_{lm}\end{aligned}$$

$$+ (\gamma_{10} + \gamma_{11})(a_{klm} + a_{kml}) + (\gamma_{12} + \gamma_{13})(a_{lkm} + a_{mkl})$$

$$+ (\gamma_{14} + \gamma_{15})(a_{lmk} + a_{mlk}) \tag{6.3.2}$$

Now, from Eq. (6.3.1), $\lambda_{klm} + \lambda_{kml} = 0$. Furthermore, by making use of Eq. (6.2.2), from which $a_{rrm} = 0$, Eq. (6.3.2) reduces to

$$0 = (\gamma_1 + \gamma_4 - \gamma_2 - \gamma_5)(a_{mrr}\delta_{kl} + a_{lrr}\delta_{km})$$

$$+ 2(\gamma_7 - \gamma_8) a_{krr}\delta_{lm}$$

$$+ (\gamma_{10} + \gamma_{11} - \gamma_{12} - \gamma_{13})(a_{klm} + a_{kml}) \tag{6.3.3}$$

Since this must hold for all motions,

$$\left.\begin{aligned} &\gamma_1 - \gamma_2 + \gamma_4 - \gamma_5 = 0 \quad , \quad \gamma_7 - \gamma_8 = 0 \\ &\gamma_{10} - \gamma_{12} + \gamma_{11} - \gamma_{13} = 0 \end{aligned}\right\} \tag{6.3.4}$$

Using this result and the fact that $a_{rri} = 0$, Eq. (5.9.20) finally reduces to

$$\lambda_{klm} = (\gamma_1 - \gamma_2)(a_{mrr}\delta_{kl} - a_{lrr}\delta_{km})$$

$$+ (\gamma_{10} - \gamma_{12})(a_{klm} - a_{kml}) + (\gamma_{14} - \gamma_{15}) a_{lmk} \tag{6.3.5}$$

Because of skew-symmetry the number of independent components of λ_{ijk} and ν_{ij} are, respectively, nine and three. They may therefore be measured by pseudotensors m_{ij} and ν_i such that

$$\left.\begin{aligned} &m_{ij} = -e_{jrs}\lambda_{irs} \quad , \quad \lambda_{ijk} = -\frac{1}{2}e_{jkr}m_{ir} \\ &\nu_i = \frac{1}{2}e_{ijk}\nu_{jk} \quad , \quad \nu_{ij} = e_{ijr}\nu_r \end{aligned}\right\} \tag{6.3.6}$$

The axial vector ν_i is called the microrotation vector, and m_{ij} is the couple stress tensor. Similarly, the microinertial rotation, $\dot{\sigma}_i$, and the body moment, l_i, may be defined by

$$\left.\begin{aligned} &\dot{\sigma}_i = -e_{irs}\dot{\sigma}_{rs} \quad , \quad \dot{\sigma}_{[ij]} = -2e_{rij}\dot{\sigma}_r \\ &l_i = -e_{irs}l_{rs} \quad , \quad l_{[ij]} = -2e_{rij}l_r \end{aligned}\right\} \tag{6.3.7}$$

A multiplication of Eq. (5.5.15) by e_{iml} gives

$$m_{rk,r} + e_{klr}t_{lr} + \rho(l_k - \dot{\sigma}_k) = 0 \tag{6.3.8}$$

Similarly, a substitution for λ_{klm} from Eq. (6.3.6) d in the energy equation, given by Eq. (5.7.14), results in

$$\rho\dot{\epsilon} = t_{kl} v_{l,k} + (s_{ml} - t_{ml})\nu_{ml} + (-\frac{1}{2} e_{lmr} m_{kr})\nu_{ml,k} - q_{k,k} + \rho h$$

which reduces to

$$\rho\dot{\epsilon} = t_{kl}(v_{l,k} - e_{klr}\nu_r) + m_{kl}\nu_{l,k} - q_{k,k} + \rho h \tag{6.3.9}$$

the term $s_{ml}\nu_{ml}$ being zero by virtue of the symmetry of s_{ml} and the skew-symmetry of ν_{ml}. An alternative useful form of this equation is obtained by using $v_{l,k} = d_{kl} + w_{kl} = d_{kl} + e_{klm}\,\omega_m$, where $w_{kl} = \frac{1}{2}(v_{l,k} - v_{k,l})$ is the classical spin tensor and ω_r the vorticity vector. Using this expression, Eq. (6.3.9) becomes

$$\rho\dot{\epsilon} = t_{kl}d_{kl} + e_{klr}t_{kl}(\omega_r - \nu_r) + m_{kl}\nu_{l,k} - q_{k,k} + \rho h \tag{6.3.10}$$

From the skew-symmetry of ν_{ij} and the definition of b_{ij}, $b_{[ij]} = \nu_{ij} - w_{ij}$, so that Eq. (5.10.1) a can be written as

$$\begin{aligned} t_{kl} &= (-\pi + \lambda_v v_{r,r} + \lambda_0 \nu_{rr})\delta_{kl} + \mu_v(v_{l,k} + v_{k,l}) \\ &\quad + \mu_0(2\nu_{kl} + v_{k,l} - v_{l,k}) + \mu_1(2\nu_{lk} + v_{l,k} - v_{k,l}) \\ &= (-\pi + \lambda_r v_{r,r})\delta_{kl} + (\mu_v + \mu_0 - \mu_1)(v_{l,k} + v_{k,l}) \\ &\quad + 2\mu_0(\nu_{kl} - v_{l,k}) + 2\mu_1(\nu_{lk} + v_{l,k}) \end{aligned} \tag{6.3.11}$$

where use has been made of $\nu_{rr} = 0$. By defining new viscosity coefficients

$$\lambda = \lambda_v \quad , \quad \mu = \mu_v \quad , \quad \kappa = \mu_1 - \mu_0 \tag{6.3.12}$$

the expression for t_{kl} becomes

$$\begin{aligned} t_{kl} &= (-\pi + \lambda v_{r,r})\delta_{kl} + (\mu - \kappa)(v_{l,k} + v_{k,l}) + 2\kappa(\nu_{lk} + v_{l,k}) \\ &= (-\pi + \lambda v_{r,r})\delta_{kl} + (\mu - \kappa)(v_{l,k} + v_{k,l}) \\ &\qquad + 2\kappa(v_{l,k} - e_{klr}\nu_r) \end{aligned} \tag{6.3.13}$$

which, upon using the definition of ω_i, becomes

$$t_{kl} = (-\pi + \lambda v_{r,r})\delta_{kl} + 2\mu d_{kl} + 2\kappa e_{klr}(\omega_r - \nu_r) \tag{6.3.14}$$

Also, with $\lambda_v = \eta_v = \lambda$ and $\mu_v = \zeta_v = \mu$, Eq. (5.10.1) b reduces to

$$s_{kl} = (-\pi + \lambda\, v_{r,r})\delta_{kl} + 2\mu\, d_{kl} \tag{6.3.15}$$

where, as before, this choice of viscosity coefficients has been made to make Stokesian fluids a subclass.

A multiplication of Eq. (6.3.5) by e_{rlm} gives

$$\begin{aligned} m_{kr} &= -e_{rlm}\lambda_{klm} \\ &= -(\gamma_1 - \gamma_2)(e_{rlm}a_{mss}\delta_{kl} - e_{rlm}a_{lss}\delta_{km}) \\ &\quad - (\gamma_{10} - \gamma_{12})(e_{rlm}a_{klm} - e_{rlm}a_{kml}) \\ &\quad - (\gamma_{14} - \gamma_{15})e_{rlm}a_{lmk} \end{aligned}$$

which, upon using the identity $e_{ijr}e_{mnr} = \delta_{im}\delta_{jn} - \delta_{in}\delta_{jm}$ and $a_{klm} = \nu_{kl,m} = e_{klr}\nu_{r,m}$, simplifies to

$$\begin{aligned} m_{kr} = 2(\gamma_1 - \gamma_2)(\nu_{r,k} - \nu_{k,r}) - 2(\gamma_{10} - \gamma_{12})(\nu_{s,r}\delta_{rk} - \nu_{r,k}) \\ - 2(\gamma_{14} - \gamma_{15})\nu_{r,k} \end{aligned}$$

This can be further reduced to

$$m_{kl} = 4\alpha\nu_{r,r}\delta_{kl} + 4\beta\nu_{k,l} + 4\gamma\nu_{l,k} \tag{6.3.16}$$

where

$$\left.\begin{aligned} \alpha = \frac{1}{2}(\gamma_{12} - \gamma_{10}) \quad , \quad \beta = \frac{1}{2}(\gamma_2 - \gamma_1) \\ \gamma = \frac{1}{2}(\gamma_1 - \gamma_2 + \gamma_{10} - \gamma_{12} - \gamma_{14} + \gamma_{15}) \end{aligned}\right\} \tag{6.3.17}$$

Substitution from Eqs. (6.3.14) and (6.3.16) in Eq. (6.3.10) gives

$$\begin{aligned} \rho\dot{\epsilon} = -\pi d_{kk} + \lambda d_{kk}d_{ll} + 2\mu d_{kl}d_{lk} + 2\kappa e_{klr}e_{kls}(\omega_r - \nu_r)(\omega_s - \nu_s) \\ + 4\alpha\nu_{r,r}\nu_{s,r} + 4\beta\nu_{k,l}\nu_{l,k} + 4\gamma\nu_{l,k}\nu_{l,k} - q_{k,k} + \rho h \end{aligned}$$

which can also be written as

$$\begin{aligned} \rho\dot{\epsilon} = -\pi d_{kk} + \lambda d_{kk}d_{ll} + 2\mu d_{kl}d_{lk} + 4\kappa(\omega_k - \nu_k)(\omega_k - \nu_k) \\ + 4\alpha\nu_{k,k}\nu_{l,l} + 4\beta\nu_{k,l}\nu_{l,k} + 4\gamma\nu_{l,k}\nu_{l,k} - q_{k,k} + \rho h \end{aligned} \tag{6.3.18}$$

The fluid will be assumed to be microisotropic. Equations (6.2.3) and (6.2.5) then yield

$$i_{km} = \frac{1}{2}j\delta_{km} \tag{6.3.19}$$

where j is constant along material lines.

Using Eqs. (5.5.16) a and (6.2.6),

$$\dot{\sigma}_r = -e_{rkl}\dot{\sigma}_{[kl]} = -ie_{rkl}\dot{\nu}_{lk} = 2i\dot{\nu}_r$$

Thus

$$\dot{\sigma}_k = j\dot{\nu}_k \tag{6.3.20}$$

In general, the balance law for stress moments, given by Eq. (5.5.15), has more information than the balance law for couple stresses given in Eq. (2.6.5) or Eq. (6.3.8). However, as shown in Section (5.5), the skew-symmetric part of Eq. (5.5.15) may be identified completely with the balance law for couple stresses. Thus, in micro fluids it is the symmetric part, given by Eq. (5.5.16) as

$$\lambda_{k(lm),k} + t_{(ml)} - s_{ml} + \rho[l_{(lm)} - \dot{\sigma}_{(lm)}] = 0 \tag{6.3.21}$$

which contains information beyond couple stresses. Now for micropolar fluids $\lambda_{k(lm)} = 0$, and from Eqs. (6.3.14) and (6.3.15) $t_{(ml)} = s_{ml}$, so that the above equation reduces to

$$\dot{\sigma}_{(lm)} = l_{(lm)} \tag{6.3.22}$$

A comparison of this equation with Eq. (6.2.6) implies that

$$\dot{\sigma}_{(lm)} = i\nu_{lr}\nu_{rm} = l_{(lm)} \tag{6.3.23}$$

In order for the balance law for first stress moments to be satisfied, the symmetric part of the first body moment must have the special form

$$l_{(lm)} = i\nu_{lr}\nu_{rm} = i(\nu_l\nu_m - \nu^2\delta_{lm}) \tag{6.3.24}$$

where $\nu^2 = \nu_r\nu_r$.

Thus, the basic equations of motion that are satisfied by micropolar fluids are the same as those for micro fluids, with some additional restrictions placed on the constitutive equations. Therefore micropolar fluids are a subclass of micro fluids so long as the symmetric part of the first body moment has the special form given in Eq. (6.3.24).

To summarize, micropolar fluids are a subclass of micro fluids that are microisotropic, satisfy $\lambda_{klm} = -\lambda_{kml}$ and $\nu_{kl} = -\nu_{lk}$, and for which the symmetric part of the first body moment has the special form given in Eq. (6.3.24).

6.4 Thermodynamics of Micropolar Fluids

Since i_{km} is a constant given by $i_{km} = i\delta_{km}$, the internal energy function may be assumed to be of the form

$$\epsilon = \epsilon(\eta, \rho^{-1}, d_{kl}, b_{kl}, a_{klm}) \tag{6.4.1}$$

This is a little more general than the expression for ϵ, for a micro fluid, given in Eq. (5.8.1).

All motions must satisfy the Clausius-Duhem inequality

$$\rho\gamma = \rho\dot{\eta} + \left[\frac{q_k}{\theta}\right]_{,k} - \frac{\rho h}{\theta} \geqslant 0 \tag{6.4.2}$$

Now for a micropolar fluid Eq. (6.3.9) may be used to eliminate $\rho h/\theta$ from Eq. (6.4.2). This results in

$$\rho(\dot{\eta} - \frac{1}{\theta}\dot{\epsilon}) + \frac{1}{\theta}t_{kl}(v_{l,k} - e_{klr}\nu_r) + \frac{1}{\theta}m_{kl}\nu_{l,k} - \frac{q_k\theta_{,k}}{\theta^2} \geqslant 0$$

which, upon use of Eq. (6.4.1), gives

$$\begin{aligned} \rho\gamma = \rho\dot{\eta}\left[1 - \frac{1}{\theta}\frac{\partial\epsilon}{\partial\eta}\right] + \frac{1}{\theta}\frac{\partial\epsilon}{\partial\rho^{-1}}\frac{\dot{\rho}}{\rho} \\ - \frac{\rho}{\theta}\left[\frac{\partial\epsilon}{\partial d_{kl}}\dot{d}_{kl} + \frac{\partial\epsilon}{\partial b_{kl}}\dot{b}_{kl} + \frac{\partial\epsilon}{\partial a_{klm}}\dot{a}_{klm}\right] \\ + \frac{1}{\theta}t_{kl}(v_{l,k} - e_{klr}\nu_r) + \frac{1}{\theta}m_{kl}\nu_{l,k} - \frac{q_k\theta_{ik}}{\theta^2} \geqslant 0 \end{aligned} \tag{6.4.3}$$

This inequality must be satisfied for all independent changes in $\dot{\eta}$, $\dot{\mathbf{D}}$, $\dot{\mathbf{B}}$ and $\theta_{,k}$. Since Eq. (6.4.3) is linear in these quantities, it cannot be maintained for all independent variations of these quantities unless

$$\frac{\partial\epsilon}{\partial d_{(kl)}} = 0\,, \quad \frac{\partial\epsilon}{\partial b_{kl}} = 0\,, \quad \frac{\partial\epsilon}{\partial a_{klm}} = 0 \tag{6.4.4}$$

$$\theta = \frac{\partial\epsilon}{\partial\eta}\Big|_{\rho^{-1}}\,, \quad q_k = 0 \tag{6.4.5}$$

$$\rho\gamma \equiv -\frac{1}{\theta}\frac{\partial\epsilon}{\partial\rho^{-1}}d_{kk} + \frac{1}{\theta}t_{kl}(v_{l,k} - e_{klr}\nu_r) + \frac{1}{\theta}m_{kl}\nu_{l,k} \geqslant 0 \tag{6.4.6}$$

where $\dot{\rho}$ has been replaced by $-\rho\, d_{kk}$. Equations (6.4.4) b and c show that ϵ

must be independent of **B** and **Λ**. Furthermore, since any function ϵ of a symmetric tensor d_{kl} can always be expressed as a function of $d_{(kl)}$, the equation

$$\frac{\partial \epsilon}{\partial d_{(kl)}} = \frac{1}{2}\left[\frac{\partial \epsilon}{\partial d_{kl}} + \frac{\partial \epsilon}{\partial d_{kl}}\right] = 0$$

shows that ϵ cannot depend on **D**.

Thus ϵ must be independent of **D**, **B** and **Λ**. Using the constitutive equations, given in Eqs. (6.3.14) and (6.3.16), the inequality in Eq. (6.4.6) can be reduced further to

$$\rho\gamma \equiv \frac{1}{\theta}[\lambda\, d_{kk} d_{ll} + 2\mu\, d_{kl} d_{lk} + 4\kappa(\omega_k - \nu_k)(\omega_k - \nu_k)$$

$$+ 4\alpha\nu_{k,k}\nu_{l,l} + 4\beta\nu_{k,l}\nu_{l,k} + 4\gamma\nu_{l,k}\nu_{l,k}] \geqslant 0 \tag{6.4.7}$$

In order for this inequality to be satisfied for all possible motions, that is, for all independent d_{ij}, ω_i, ν_i and $\nu_{i,j}$,

$$\lambda\, d_{kk} d_{ll} + 2\mu\, d_{kl} d_{lk} \geqslant 0 \tag{6.4.8}$$

$$\kappa(\omega_k - \nu_k)(\omega_k - \nu_k) \geqslant 0 \tag{6.4.9}$$

$$\alpha\nu_{k,k}\nu_{l,l} + \beta\nu_{k,l}\nu_{l,k} + \gamma\nu_{l,k}\nu_{l,k} \geqslant 0 \tag{6.4.10}$$

where it has been assumed that $\theta > 0$. The necessary and sufficient conditions for these three inequalities to hold are given, respectively, by

$$3\lambda + 2\mu \geqslant 0 \quad , \quad \mu \geqslant 0 \tag{6.4.11}$$

$$\kappa \geqslant 0 \tag{6.4.12}$$

$$\gamma - \beta \geqslant 0\,,\ \gamma + \beta \geqslant 0\,,\ 3\alpha + \beta + \gamma \geqslant 0 \tag{6.4.13}$$

These inequalities, relating the material constants λ, μ, κ, α, β and γ, can be regrouped as

$$\left.\begin{array}{l} 3\lambda + 2\mu \geqslant 0\,,\ \mu \geqslant 0\,,\ \kappa \geqslant 0 \\ 3\alpha + 2\gamma \geqslant 0\,,\ -\gamma \leqslant \beta \leqslant \gamma \quad , \quad \gamma \geqslant 0 \end{array}\right\} \tag{6.4.14}$$

6.5 Equations of Motion

Micropolar fluids obviously satisfy the continuity equation

$$\dot{\rho} + \rho v_{r,r} = 0 \tag{6.5.1}$$

A substitution from Eqs. (6.3.13) and (6.3.16) in Eqs. (5.5.8) and (6.3.8) gives the equations of motion

$$\rho a_k = -\pi_{,k} + \lambda v_{r,rk} + (\mu - \kappa)(v_{r,kr} + v_{k,rr}) + 2\kappa(v_{k,rr} - e_{rks}\nu_{s,r}) + \rho f_k$$

$$\rho \dot{\sigma}_k = 4\alpha\nu_{r,rk} + 4\beta\nu_{k,kr} + 4\gamma\nu_{k,rr} + e_{krs}[(-\pi + \lambda v_{t,t})\delta_{rs} + (\mu - \kappa)d_{rs} + 2\kappa(v_{s,r} - e_{rst}\nu_t)] + \rho l_k$$

which, on simplification and use of Eq. (6.3.20), reduce to

$$\rho a_k = -\pi_{,k} + (\lambda + \mu - \kappa)(v_{r,r})_{,k} + (\mu + \kappa)v_{k,rr} + 2\kappa e_{krs}\nu_{s,r} + \rho f_k \tag{6.5.2}$$

$$\rho j \dot{\nu}_k = 4(\alpha + \beta)(\nu_{r,r})_{,k} + 4\gamma\nu_{k,rr} + 2\kappa e_{krs} v_{s,r} - 4\kappa\nu_k + \rho l_k \tag{6.5.3}$$

In Gibbsian vector notation these equations may be written as

$$\rho \mathbf{a} = -\nabla\pi + (\lambda + \mu - \kappa)\nabla\nabla\cdot\mathbf{v} + (\mu + \kappa)\nabla^2\mathbf{v} + 2\kappa\nabla\times\boldsymbol{\nu} + \rho\mathbf{f} \tag{6.5.4}$$

$$\rho j \dot{\boldsymbol{\nu}} = 4(\alpha + \beta)\nabla\nabla\cdot\boldsymbol{\nu} + 4\gamma\nabla^2\boldsymbol{\nu} + 2\kappa\nabla\times\mathbf{v} - 4\kappa\boldsymbol{\nu} + \rho\mathbf{l} \tag{6.5.5}$$

or in the alternative forms:

$$\rho \mathbf{a} = -\nabla\pi + (\lambda + 2\mu)\nabla\nabla\cdot\mathbf{v} - (\mu + \kappa)\nabla\times\nabla\times\mathbf{v} + 2\kappa\nabla\times\boldsymbol{\nu} + \rho\mathbf{f} \tag{6.5.6}$$

$$\rho j \dot{\boldsymbol{\nu}} = 4(\alpha + \beta + \gamma)\nabla\nabla\cdot\boldsymbol{\nu} - 4\gamma\nabla\times\nabla\times\boldsymbol{\nu} + 2\kappa\nabla\times\mathbf{v} - 4\kappa\boldsymbol{\nu} + \rho\mathbf{l} \tag{6.5.7}$$

Together with the continuity equation, Eqs. (6.5.2) and (6.5.3) give seven equations in the seven unknowns: ρ, $\mathbf{v}$ and $\boldsymbol{\nu}$.

6.6 Boundary and Initial Conditions

The initial conditions may be of the type

$$\rho(\mathbf{x},0) = \rho_0(\mathbf{x}) \ , \quad v_k(\mathbf{x},0) = v_{0k}(\mathbf{x}) \ , \quad \nu_k(\mathbf{x},0) = \nu_{0k}(\mathbf{x}) \tag{6.6.1}$$

The boundary conditions that are used are of the type

$$\mathbf{v}(\mathbf{x}_B,t) = \mathbf{v}_B \ , \quad \boldsymbol{\nu}(\mathbf{x}_B,t) = \boldsymbol{\nu}_B \tag{6.6.2}$$

Instead of these boundary conditions, the forces and moments may be prescribed at the boundary, in which case

$$t_{kl} n_k = t_l \quad , \quad m_{kl} n_k = m_l \tag{6.6.3}$$

at the boundary, where t_l and m_l are the imposed stress vector and the couple stress vector, respectively.

Other types of mixed conditions are also possible.

6.7 Two Limiting Cases

There are two interesting limiting cases of micropolar fluids: (i) microstructure absent, but couple stresses present, and (ii) couple stresses absent, but microstructure present.

If the microrotation is constrained in such as way that $\boldsymbol{\nu} = \boldsymbol{\omega}$, then the entire kinematics of flow is determined by the velocity field, since the microstructure is absent. This leads to the following cases:

A. The Case of Constrained Microrotation

The equations for a micropolar fluid may be summarized, from Eqs. (6.3.13), (6.3.16), (6.5.2) and (6.5.3), as

$$t_{kl} = (-\pi + \lambda d_{rr})\delta_{kl} + 2\mu d_{kl} + 2\kappa e_{klr}(\omega_r - \nu_r) \tag{6.7.1}$$

$$m_{kl} = 4\alpha \nu_{r,r}\delta_{kl} + 4\beta \nu_{k,l} + 4\gamma \nu_{l,k} \tag{6.7.2}$$

$$\rho a_k = -\pi_{,k} + (\lambda + \mu - \kappa)(v_{r,r})_{,k} + (\mu + \kappa) v_{k,rr} + 2\kappa e_{krs}\nu_{s,r} + \rho f_k \tag{6.7.3}$$

$$\rho \dot{\sigma}_k = \rho j \dot{\nu}_k = 4(\alpha + \beta)(\nu_{r,r})_{,k} + 4\gamma \nu_{k,rr} + 2\kappa e_{krs} v_{s,r} - 4\kappa \nu_k + \rho l_k \tag{6.7.4}$$

By taking the curl of Eq. (6.7.4)

$$4\kappa e_{krs}\nu_{s,r} = 4\gamma e_{krs}\nu_{s,mmr} + 2\kappa e_{krs}e_{smn}v_{n,mr} + e_{krs}(\rho l_s - \rho\dot{\sigma}_s)_{,r}$$

which, on simplification, reduces to

$$2\kappa e_{krs}\nu_{s,r} = 2\gamma e_{krs}\nu_{s,rmm} + \kappa(v_{r,kr} - v_{k,rr}) + \frac{1}{2}e_{krs}(\rho l_s - \rho\dot{\sigma}_s)_{,r} \tag{6.7.5}$$

Use of this expression in Eq. (6.7.3) gives

$$\rho a_k = -\pi_{,k} + (\lambda + \mu)(v_{r,r})_{,k} + \mu v_{k,rr} + 2\gamma e_{krs}\nu_{s,rmm} + \rho f_k + \frac{1}{2}e_{krs}(\rho l_s - \rho\dot{\sigma}_s)_{,r} \tag{6.7.6}$$

If the microrotation is constrained such that $\boldsymbol{\nu} = \boldsymbol{\omega}$, so that there is really no microstructure, $\dot{\boldsymbol{\sigma}}$ may be taken to be zero. Also $\nu_{r,r} = \omega_{r,r} = 0$, $\nu_{k,l} = \omega_{k,l} = k_{lk}$, $\nu_{l,k} = k_{kl}$, and $e_{krs}\nu_{s,rmm} = e_{krs}\omega_{s,rmm} = \frac{1}{2}e_{krs}e_{spq}v_{q,prmm} = \frac{1}{2}(v_{r,rkmm} - v_{k,rrmm})$, so that Eqs. (6.7.1), (6.7.2) and (6.7.3) reduce to

$$t_{kl} = t_{(kl)} = (-\pi + \lambda d_{rr})\delta_{kl} + 2\mu d_{kl} \tag{6.7.7}$$

$$m_{kl} = m_{kl}^D = 4\gamma k_{kl} + 4\beta k_{lk} \tag{6.7.8}$$

$$\rho a_k = -\pi_{,k} + (\lambda + \mu)(v_{r,r})_{,r} + \gamma(v_{r,r})_{,kss} + \mu v_{k,rr} - \gamma v_{k,rrss} + \rho f_k + \frac{1}{2}e_{krs}(\rho l_s)_{,r} \tag{6.7.9}$$

These three equations are the same as Eq. (3.2.3), (3.2.4) and (3.3.6), respectively, with $\gamma = \eta$. Thus, the theory of couple stresses in fluids, discussed in Chapter 3, is a special case of micropolar fluids with the microrotations constrained such that $\boldsymbol{\nu} = \boldsymbol{\omega}$, and may therefore be considered as the appropriate theory of couple stresses in fluids in the absence of microstructure.

B. The Case with Couple Stresses Absent

For couple stresses to be absent, $\alpha = \beta = \gamma = 0$, so that Eqs. (6.7.1) to (6.7.4) reduce to

$$t_{kl} = (-\pi + \lambda d_{rr})\delta_{kl} + 2\mu d_{kl} + 2\kappa e_{klr}(\omega_r - \nu_r) \tag{6.7.10}$$

$$\rho a_k = -\pi_{,k} + (\lambda + \mu - \kappa)(v_{r,r})_{,k} + (\mu + \kappa) v_{k,rr}$$

$$+ 2\kappa e_{krs} \nu_{s,r} + \rho f_k \tag{6.7.11}$$

$$\rho \dot{\sigma}_k = \rho j \dot{\nu}_k = 2\kappa e_{krs} v_{s,r} - 4\kappa \nu_k + \rho l_k$$

$$= 4\kappa(\omega_k - \nu_k) + \rho l_k \tag{6.7.12}$$

These equations then describe a fluid, having zero couple stresses, in which the microstructure is such that the kinematics of motion is described by the velocity field $\mathbf{v}$ and an independent pseudovector $\boldsymbol{\nu}$ — the type of situation described by Ericksen's theory of anisotropic fluids given in Chapter 4. However, a comparison of Eq. (6.7.7) with Eq. (4.7.7) shows that Ericksen's theory is not a special case of micropolar fluids with couple stresses absent.

6.8 Steady Flow Between Parallel Plates

Only steady plane flows of incompressible fluids will be considered in this section. If the flow field is assumed to have the form

$$\mathbf{v} = \begin{bmatrix} v_x \\ v_y \\ v_z \end{bmatrix} = \begin{bmatrix} u(y) \\ 0 \\ 0 \end{bmatrix} , \quad \boldsymbol{\nu} = \begin{bmatrix} \nu_x \\ \nu_y \\ \nu_z \end{bmatrix} = \begin{bmatrix} 0 \\ 0 \\ \nu(y) \end{bmatrix} \tag{6.8.1}$$

then the equations of motion, given by Eqs. (6.5.2) and (6.5.3), reduce to

$$\left.\begin{aligned} &(\mu + \kappa)\frac{d^2u}{dy^2} + 2\kappa\frac{d\nu}{dy} - \frac{dp}{dx} = 0 \ , \quad p = p(x) \\ &4\gamma\frac{d^2\nu}{dy^2} - 2\kappa\frac{du}{dy} - 4\kappa\nu = 0 \end{aligned}\right\} \tag{6.8.2}$$

where the hydrostatic pressure π has been denoted by p. The general solution of these equations is

$$u = -\frac{2\kappa l}{\mu + \kappa}\Big[A \sinh(y/l) + B \cosh(y/l)\Big]$$

$$+ \frac{1}{2\mu}\left[\left(\frac{dp}{dx}\right)^2 y^2 + 2Cy\right] + D \tag{6.8.3}$$

$$\nu = A \cosh(y/l) + B \sinh(y/l) - \frac{1}{2\mu}\left[\left(\frac{dp}{dx}\right) y + C\right] \tag{6.8.4}$$

where

$$l^2 = \frac{\gamma}{\mu}\left(\frac{\mu+\kappa}{\kappa}\right) \tag{6.8.5}$$

Then

$$t_{yx} = (\mu+\kappa)\frac{du}{dy} + 2\kappa\nu = \left(\frac{dp}{dx}\right)y + C \tag{6.8.6}$$

$$m_{yz} = 4\gamma\frac{d\nu}{dy} = \frac{4\gamma}{l}[A\sinh(y/l) + B\cosh(y/l)] - \frac{2\gamma}{\mu}\frac{dp}{dx} \tag{6.8.7}$$

1. Couette Flow with dp/dx = 0

Consider steady Couette flow between two parallel plates, the one at $y = 0$ being fixed, and the one at $y = h$ moving at a constant velocity V. If the fluid is assumed to adhere to the boundary, then one set of appropriate boundary conditions is

$$\left.\begin{aligned} u(0) &= 0\,, \quad u(h) = V \\ \nu(0) &= 0\,, \quad \nu(h) = 0 \end{aligned}\right\} \tag{6.8.8}$$

Under these conditions, the general solution reduces to

$$\frac{u}{V} = \frac{1}{D(a)}\left[a\xi\sinh a - \frac{\kappa}{\mu+\kappa}\left\{\cosh a + \cosh a\xi - \cosh a(1-\xi) - 1\right\}\right] \tag{6.8.9}$$

$$\frac{\nu}{(-V/2h)} = \frac{a}{D(a)}[\sinh a - \sinh a\xi - \sinh a(1-\xi)] \tag{6.8.10}$$

$$\frac{m_{yz}}{2\mu V[\kappa/(\mu+\kappa)]} = \frac{1}{D(a)}[\cosh a\xi - \cosh a(1-\xi)] \tag{6.8.11}$$

where $a = h/l$, $\xi = y/h$ and $D(a) = a\sinh a - \dfrac{2\kappa}{\mu+\kappa}(\cosh a - 1)$.

A comparison of Eqs. (6.8.9) to (6.8.11) with Eqs. (3.5.12) to (3.5.14), respectively, shows that the results for the couple stress theory for boundary conditions B, are a special case of those for micropolar fluids, and can be obtained by putting $\kappa/(\mu+\kappa) = 1$ and $\nu_z = \omega_z$ in Eqs. (6.8.9) to (6.8.11).

If instead of using the boundary conditions given in Eq. (6.8.8) the zero couple stress boundary conditions

$$\left.\begin{aligned} u(0) &= 0\,, \quad u(h) = V \\ m_{yz}(0) &= 0\,, \quad m_{yz}(h) = 0 \end{aligned}\right\} \tag{6.8.12}$$

are used, then the general solution reduces to the classical nonpolar solution

$$\frac{u}{V} = \frac{y}{h}\,, \quad \frac{\omega_z}{(-V/2h)} = 1\,, \quad m_{yz} \equiv 0 \tag{6.8.13}$$

2. Plane Poiseuille Flow

Consider the flow between two parallel plates, which are at a distance $2h$ from each other, due to a pressure gradient $-dp/dx$. Let the x axis coincide with the centerline, so that the appropriate boundary conditions are

$$u(\pm\, h) = 0 \quad , \quad \nu(\pm\, h) = 0 \tag{6.8.14}$$

The general solution then reduces to

$$U(\xi) = \frac{u(\xi)}{u_0} = 1 - \xi^2 - \left(\frac{2\kappa}{\mu + \kappa}\right)\frac{\cosh a}{a \sinh a}\left[1 - \frac{\cosh a\xi}{\cosh a}\right] \tag{6.8.15}$$

$$N(\xi) = \frac{\nu(\xi)}{(u_0/h)} = \xi - \frac{\sinh a\xi}{\sinh a} \tag{6.8.16}$$

$$M(\xi) = \frac{m_{yz}(\xi)}{[4\mu\kappa/(\mu + \kappa)]u_0} = \frac{1}{a^2}\left[1 - \frac{a \cosh a\xi}{\sinh a}\right] \tag{6.8.17}$$

where $u_0 = -\,(h^2/2\mu)\,dp/dx$.

The volumetric flow per unit depth, $Q = 2\int_0^h u(y)\,dy$, is given by

$$\frac{Q}{Q_0} = 1 - \left(\frac{\kappa}{\mu + \kappa}\right)\frac{3}{a^2}\left[a \coth a - 1\right] \tag{6.8.18}$$

where $Q_0 = -\,(2h^3/3\mu)\,dp/dx$.

The shear stress t_{yx} is given by

$$t_{yx} = \frac{dp}{dx}\,y \tag{6.8.19}$$

Here again, the solution for the couple stress theory discussed in Chapter 3 may be obtained by putting $\kappa/(\mu + \kappa) = 1$ and $\nu_z = \omega_z$, in which case Eqs. (6.8.15) to (6.8.18) reduce, respectively, to Eqs. (3.5.25) to (3.5.28).

For the problem under discussion, it follows from Eqs. (6.3.13) and (6.3.16) that (i) couple stresses will be zero if $\gamma = 0$, irrespective of the variation of v_i and ν_i and (ii) if $\kappa = 0$, then the stress will not be affected by the microrotation and the couple stress will not be affected by the velocity field. In fact, because of this uncoupling when $\kappa = 0$, the macroscopic velocity field will be the same as in the nonpolar case. Thus, the value of κ is a measure of the effect of the microstructure on the macroscopic velocity and stress fields.

The expression for l^2, given in Eq. (6.8.5), can be written as

$$l^2 = \frac{\gamma}{\mu}\left[\frac{1 + \kappa/\mu}{\kappa/\mu}\right] \tag{6.8.20}$$

Even though all the effects are present simultaneously, the motion may be assumed to be affected by (i) viscous action, which is measured by μ, (ii) the effect of couple stresses, measured by γ, and (iii) the direct coupling of the microstructure to the velocity field, measured by κ. Equation (6.4.14) shows that each of the constants μ, κ and γ can have any value greater than, or equal to, zero, so that the ratios γ/μ and κ/μ, which are measures, respectively, of the relative strengths of the couple stress to the viscous affects and the microstructure coupling to the viscous effects, can have any values greater than, or equal to, zero.

When the viscous effects are much larger than the couple stress effects, γ/μ is small, becoming zero when $\gamma = 0$. Thus in the absence of couple stresses $l \to 0$. On the other hand $l \to \infty$ either (i) when the couple stress effects are much larger than the viscous effects, that is, when couple stresses are infinite, $\gamma/\mu \to \infty$, or (ii) when $\kappa/\mu \to 0$, that is, when the coupling effects of the microstructure are negligible in comparison to the viscous effects. Thus, whereas the nondimensional parameter $a = h/l$ can be infinite only for vanishing couple stresses, for which $\gamma/\mu \to 0$, a can be zero either (i) when the couple stresses are infinite or (ii) when the effects of microstructure are negligible.

As $a \to \infty$, the limiting values of the flow parameters are given by $U(\xi) = 1 - \xi^2$, $N(\xi) = \xi$, $M(\xi) \equiv 0$, and $Q/Q_0 = 1$, so that the solution reduces to the classical nonpolar solution with $\omega_z(\xi) = \nu(\xi)$. Also, as $a \to 0$,

$$U(\xi) = \frac{\mu}{\mu + \kappa}(1 - \xi^2) \tag{6.8.21}$$

$$\frac{Q}{Q_0} = \frac{\mu}{\mu + \kappa} \tag{6.8.22}$$

so that $U(\xi) \equiv 0$ and $Q/Q_0 \equiv 0$ in the special case when $\kappa/\mu \to \infty$, that is, when the effect of microstructure coupling is much larger than the viscous effect. Thus there cannot be any flow when the couple stresses and the microstructure coupling are both infinite in comparison to the viscous effects.

The variations of the velocity profile with a, for $\kappa/\mu = \infty$ and $\kappa/\mu = 10$, are shown in Fig. 6.8.1. The profile for $\kappa/\mu = \infty$ and $a = 0$ is $U(\xi) \equiv 0$. The velocity profiles for $\kappa/\mu = 5$ and $\kappa/\mu = 1$ are shown in Fig. 6.8.2. Thus, when κ/μ is finite, the velocity field is not identically zero for any value of a. The variations of the velocity profiles with κ/μ, for $a = 1$ and $a = 10$ are shown in Fig. 6.8.3. The maximum velocity U_0 occurs at $\xi = 0$ and its variation with a, for different values of κ/μ, is shown in Fig. 6.8.4. When $\kappa/\mu = 0$, $U_0 = 1$ for all values of a. Finally, the variation of the throughflow Q/Q_0 is shown in Fig. 6.8.5, wherein $Q/Q_0 \equiv 1$ for $\kappa/\mu = 0$.

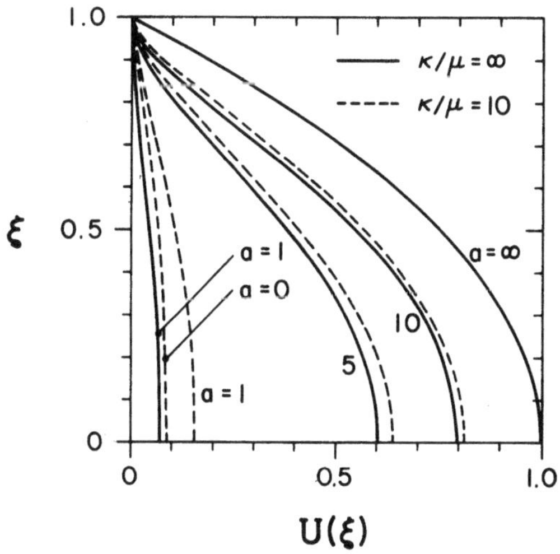

Fig. 6.8.1 Variation of the velocity $U(\xi)$ with a for $\kappa/\mu = \infty$ and $\kappa/\mu = 10$.

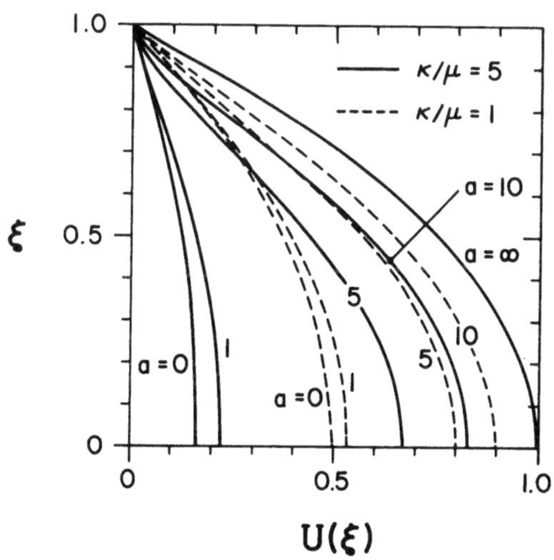

Fig. 6.8.2 Variation of the velocity $U(\xi)$ with a for $\kappa/\mu = 5$ and $\kappa/\mu = 1$.

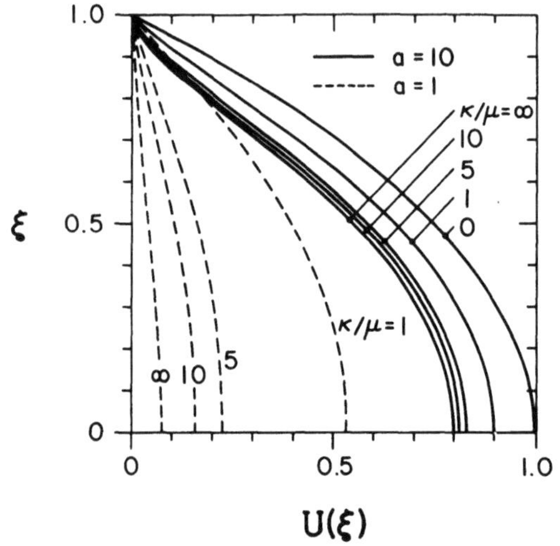

Fig. 6.8.3 Variation of the velocity $U(\xi)$ with κ/μ for $a = 1$ and $l = 10$.

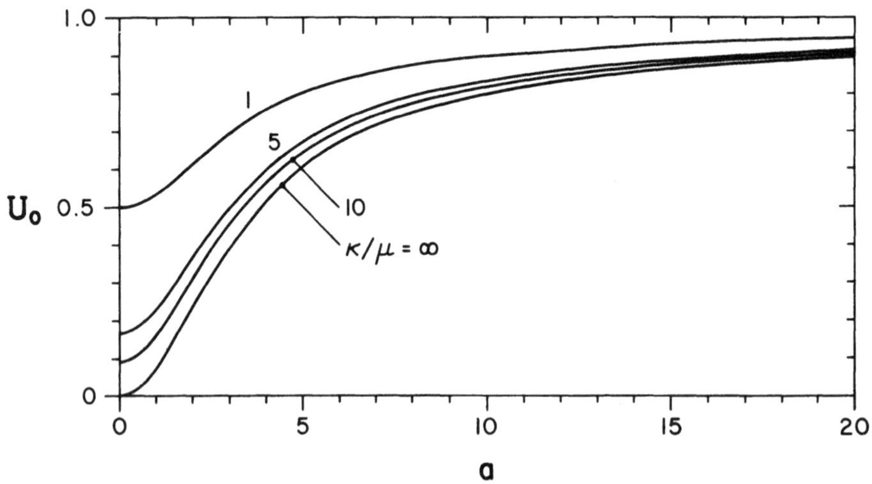

Fig. 6.8.4 Variation of the maximum velocity U_0 with a for different values of κ/μ.

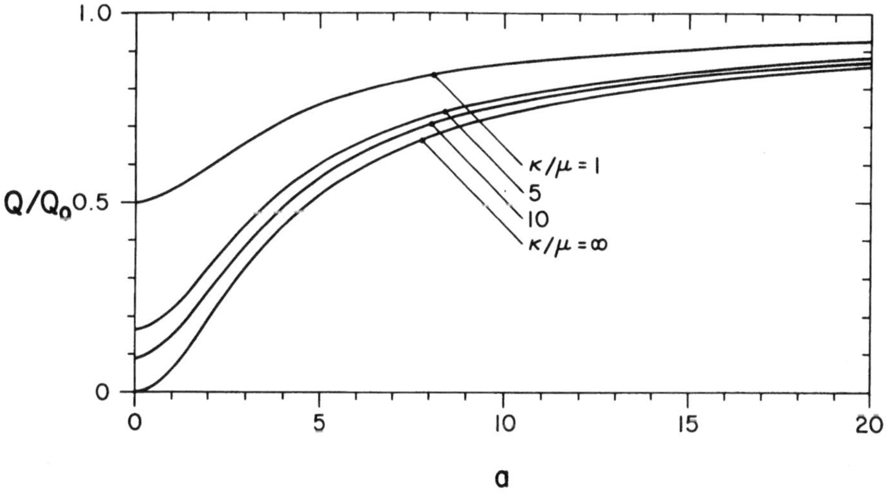

Fig. 6.8.5 Variation of the through flow Q/Q_0 with a for different values of κ/μ.

6.9 Steady Couette Flow Between Two Coaxial Cylinders

Consider the steady flow between two concentric cylinders, generated by the steady rotation of the outer cylinder with an angular velocity Ω. In cylindrical polar coordinates (r,θ,z) with the inner and outer cylinder walls at $r = \lambda R$ and $r = R$ respectively, the appropriate velocity and microrotation fields are of the form

$$\left.\begin{array}{lll} v_r = 0 \;, & v_z = 0 \;, & v_\theta = u(r) \\ \nu_r = 0 \;, & \nu_\theta = 0 \;, & \nu_z = \nu(r) \end{array}\right\} \tag{6.9.1}$$

For this case, the equations of motion, expressed in cylindrical polar coordinates, reduce to

$$\left.\begin{array}{l} (\mu + \kappa)\left(u'' + \dfrac{1}{r}u' - \dfrac{1}{r^2}u\right) - 2\kappa\nu' = 0 \\ 4\gamma(\nu'' + \dfrac{1}{r}\nu') + 2\dfrac{\kappa}{r}(ur)' - 4\kappa\nu = 0 \end{array}\right\} \tag{6.9.2}$$

The general solution of these equations is

$$u = \frac{A_1}{r} + A_2 r + A_3 I_1(r/l) + A_4 K_1(r/l) \tag{6.9.3}$$

$$\nu = A_2 + \frac{\mu + \kappa}{2\kappa l}\,[A_3 I_0(r/l) - A_4 K_0(r/l)] \tag{6.9.4}$$

where, again, $l^2 = (\gamma/\mu)(1 + \mu/\kappa)$.

Ariman, et al. [Ref. 5] used the boundary conditions:

$$u = 0\,,\ \ \nu = 0 \quad at \quad r = \lambda R\,,\ \ \lambda < 1$$

$$u = \Omega R\,,\ \ \nu = 0 \quad at \quad r = R$$

However, as pointed out by Cowin and Pennington [Ref. 6], the physically more appropriate boundary conditions are

$$\left.\begin{aligned} &u = 0\,,\ \ \nu = 0 \quad at \quad r = \lambda R\,,\ \ \lambda < 1 \\ &u = \Omega R\,,\ \ \nu = \Omega \quad at \quad r = R \end{aligned}\right\} \tag{6.9.5}$$

The conditions $u = \Omega R$ and $\nu = \Omega$ at $r = R$ ensure that, at the wall $r = R$, a fluid particle has no motion relative to the wall. If the condition $\nu = 0$ at $r = R$ is used, then a fluid particle would have an angular velocity relative to the wall.

Subject to the boundary conditions in Eq. (6.9.5), the solution is given by

$$\left.\begin{aligned} A_1 &= R\Omega\left(\frac{\lambda R}{D}\right)\left[Z_3\left\{I_1(\lambda a) - \lambda\frac{\alpha\beta}{2}I_0(\lambda a)\right\}\right. \\ &\qquad \left. - Y_3\left\{K_1(\lambda a) + \lambda\frac{\alpha\beta}{2}K_1(\lambda a)\right\}\right] \\ A_2 &= R\Omega\left(\frac{a\beta}{2RD}\right)\left[Z_3 I_0(\lambda a) + Y_3 K_0(\lambda a)\right] \\ A_3 &= -\frac{R\Omega}{D}\,Z_3\,,\quad A_4 = \frac{R\Omega}{D}\,Y_3 \end{aligned}\right\} \tag{6.9.6}$$

where

$$\beta = (\mu + \kappa)/\kappa$$

$$\left.\begin{aligned} Y_n &= I_n(a) - \lambda^n I_n(\lambda a) \\ Z_n &= K_n(a) - \lambda^n K_n^{\cdot}](\lambda a) \end{aligned}\right\} n = 0,\, 1$$

$$Y_2 = Y_1 - (1 - \lambda^2)\frac{a\beta}{2}I_0(\lambda a)$$

$$Z_2 = Z_1 + (1 - \lambda^2)\frac{a\beta}{2}K_0(\lambda a)$$

$$Y_3 = Y_0 - \frac{2}{a\beta} Y_2 , \quad Z_3 = Z_0 - \frac{2}{a\beta} Z_2$$

$$D = Y_0 Z_2 + Y_2 Z_0$$

Here again, with $\kappa/(\mu + \kappa) = 1$ and $\nu_z = \omega_z$, the above solution reduces to the couple stress solution given in Eq. (3.6.13).

6.10 Pipe Poiseuille Flow

Consider the steady flow in a pipe of radius R. Then in cylindrical polar coordinates, (r,θ,z), the velocity $\mathbf{v}$ and the microrotation $\boldsymbol{\nu}$ may be taken to be

$$\left.\begin{aligned} v_r = v_\theta \equiv 0 , \quad v_z = u(r) \\ \nu_r = \nu_z \equiv 0 , \quad \nu_\theta = \nu(r) \end{aligned}\right\} \tag{6.10.1}$$

For this case the equations of motion, in cylindrical polar coordinates, reduce to

$$0 = -\frac{dp}{dz} + (\mu + \kappa)\left[\frac{1}{r}\frac{d}{dr}\left(r\frac{du}{dr}\right) + \frac{2\kappa}{r}\frac{d}{dr}(r\nu)\right] , \quad p = p(z) \tag{6.10.2}$$

$$0 = 4\gamma\frac{d}{dr}\left[\frac{1}{r}\frac{d}{dr}(r\nu)\right] - 2\kappa\frac{du}{dr} - 4\kappa\nu \tag{6.10.3}$$

The general solution of Eqs. (6.10.2) and (6.10.3) is given by

$$u = \frac{1}{4\mu}\frac{dp}{dz} r^2 + A_1 + A_2 \ln r + A_3 I_0(r/l) + A_4 K_0(r/l) \tag{6.10.4}$$

$$\nu = -\frac{1}{4\mu}\frac{dp}{dz} r + \frac{1}{2}\left(\frac{\mu + \kappa}{\mu - \kappa}\right)\frac{A_2}{r}$$

$$+ \frac{\mu + \kappa}{2\kappa l}\left[-A_3 I_1(r/l) + A_4 K_1(r/l)\right] \tag{6.10.5}$$

The appropriate boundary conditions are $u(R) = 0$, $\nu(R) = 0$ and u and ν be finite at $r = 0$. The latter set of conditions requires that $A_2 = A_4 = 0$. Subject to these boundary conditions, the general solution reduces to

$$U(\xi) = \frac{u(\xi)}{u_0} = 1 - \xi^2 - \left(\frac{2\kappa}{\mu + \kappa}\right)\frac{1}{a}\frac{I_0(a)}{I_1(a)}\left[1 - \frac{I_0(a\xi)}{I_0(a)}\right] \tag{6.10.6}$$

$$N(\xi) = \frac{R\nu(\xi)}{u_0} = \xi - \frac{I_1(a\xi)}{I_1(a)} \tag{6.10.7}$$

where $l^2 = (\gamma/\mu)(1 + \kappa/\mu)$ and $a = R/l$.

As $a \to \infty$, this solution tends to the classical nonpolar solution $U(\xi) = 1 - \xi^2$. Also, as $a \to 0$,

$$U(\xi) = \frac{\mu}{\mu + \kappa}(1 - \xi^2) \tag{6.10.8}$$

so that for the special case of $a \to 0$ and $\mu/\kappa \to 0$

$$U(\xi) \equiv 0 \tag{6.10.9}$$

As in earlier cases, the solution given by Eqs. (6.10.6) and (6.10.7) reduces to the couple stress solution, given in Eqs. (3.7.15) and (3.7.16), when $\nu = \omega$ and $\kappa/(\mu + \kappa) = 1$.

6.11 Micropolar Fluids with Stretch

In a microfluid the gyration tensor $\mathbf{N} = (\nu_{ij})$ has nine independent components. The assumption of skew-symmetry of $\mathbf{N}$ led to the theory of micropolar fluid for which the micromotion consists purely of rotations about the centroid of a macroelement, no microdeformations being admissible.

If $\mathbf{N}$ is assumed to have the form

$$\nu_{kl} = n\delta_{kl} + e_{klr}\nu_r \tag{6.11.1}$$

then

$$\nu_{(kl)} = n\delta_{kl} \quad , \quad \nu_{[kl]} = e_{klr}\nu_r \tag{6.11.2}$$

so that if only motions about the centroid of a macroelement are considered, for which $dx_k = 0$, then from Eq. (6.2.1)

$$\frac{D}{Dt}(ds')^2 = 2\nu_{(kl)}d\xi_k d\xi_l = 2nd\xi_r d\xi_r = 2n(ds')^2$$

Thus, for a fluid for which $\mathbf{N}$ has the form given in Eq. (6.11.1), besides microrotations, the micromotion also allows for microstretch and shear.

A microfluid will be called a microstretch fluid, or a micropolar fluid with stretch if, for all motions, the gyration and first stress moment tensors have the form

$$\nu_{kl} = n\delta_{kl} + e_{klr}\nu_r \quad , \quad \lambda_{klm} = \lambda_k\delta_{lm} - \frac{1}{2}e_{lmr}m_{kr} \tag{6.11.3}$$

The difference between these fluids and micropolar fluids is due to the presence of nonzero n and λ_k. The presence of n allows microelements to stretch. That microstretch fluids exist as a subclass of microfluids will now be shown.

If in addition to Eq. (6.11.1), the fluid is assumed to be microisotropic, so that $i_{km} = i\delta_{km} = \frac{1}{2}j\delta_{km}$, then Eq. (5.6.7) gives $\frac{D}{Dt}(i_{km}) = 2i\nu_{(km)} = 2in\delta_{km}$, or

$$\frac{D}{Dt}(j) - 2nj = 0 \tag{6.11.4}$$

Thus, in contrast to micropolar fluids, j is not a constant along material lines.

From Eqs. (5.11.6) and (6.11.1)

$$\dot{\sigma}_{(kl)} = \frac{1}{2}j\left[(\dot{n} + n^2 - \nu_r\nu_r)\delta_{kl} + \nu_k\nu_l\right] \tag{6.11.5}$$

$$\dot{\sigma}_{[kl]} = -\frac{1}{2}je_{klr}(\dot{\nu}_r + 2n\nu_r) \tag{6.11.6}$$

Substituting from Eq. (6.11.1) in Eq. (5.10.1), the stress tensor is given by

$$t_{kl} = (-\pi + \lambda v_{r,r} + \lambda_1 n)\delta_{kl} + 2\mu d_{kl} + 2\kappa e_{klr}(\omega_r - \nu_r) \tag{6.11.7}$$

where

$$\lambda = \lambda_v , \mu = \mu_v , \kappa = \mu_1 - \mu_0 , \lambda_1 = 3\lambda_0 + 2\mu_0 + 2\mu_1 \tag{6.11.8}$$

Also, the expression for the microstress average tensor s_{kl}, given in Eq. (5.10.2), reduces to

$$s_{kl} = (-\pi + \lambda d_{rr} + \eta_1 n)\delta_{kl} + 2\mu d_{kl} \tag{6.11.9}$$

where η_1 is defined by

$$\eta_1 = 3\eta_0 + 2\zeta_1 \tag{6.11.10}$$

and, in order to include Stokesian fluids as a subclass, the viscosity coefficients have been taken as

$$\lambda = \eta_v \quad , \quad \mu = \zeta_v \tag{6.11.11}$$

Substitution from Eqs. (6.11.1) in Eq. (5.9.20) results in

$$\lambda_{k(lm)} = (\alpha_0 n_{,k} + \alpha_1 e_{krs}\nu_{s,r})\delta_{lm} + (\alpha_2 n_{,l} + \alpha_3 e_{lrs}\nu_{s,r})\delta_{km}$$

$$+ (\alpha_2 n_{,m} + \alpha_3 e'_{mrs}\nu_{s,r})\delta_{kl}$$

$$+ \alpha_4(e_{klr}\nu_{r,m} + e_{kmr}\nu_{r,l}) \tag{6.11.12}$$

$$\lambda_{k[lm]} = -2\alpha(e_{klr}\nu_{r,m} - e_{kmr}\nu_{r,l}) - 2(\beta_0 n_{,l} - \beta e_{lrs}\nu_{s,r})\delta_{km}$$

$$+ 2(\beta_0 n_{,m} - \beta e_{mrs}\nu_{s,r})\delta_{kl} - 2(\alpha + \beta + \gamma)e_{lmr}\nu_{r,k} \tag{6.11.13}$$

where

$$\alpha_1 = \gamma_7 - \gamma_8$$

$$\alpha_2 = \frac{1}{2}(\gamma_1 + \gamma_2 + 3\gamma_3 + \gamma_4 + \gamma_5 + 3\gamma_6 + \gamma_{10} + \gamma_{11} + \gamma_{12} + \gamma_{13})$$

$$\alpha_3 = \frac{1}{2}(\gamma_1 - \gamma_2 + \gamma_4 - \gamma_5)\ , \quad \alpha_4 = \frac{1}{2}(\gamma_{10} + \gamma_{11} - \gamma_{12} - \gamma_{13})$$

$$4\beta_0 = \gamma_1 + \gamma_2 + 3\gamma_3 - \gamma_4 - \gamma_5 - 3\gamma_6 + \gamma_{10} - \gamma_{11} + \gamma_{12} + \gamma_{13}$$

$$4\alpha = -\gamma_{10} + \gamma_{11} + \gamma_{12} - \gamma_{13}\ , \quad 4\beta = -\gamma_1 + \gamma_2 + \gamma_4 - \gamma_5$$

$$4\gamma = \gamma_1 - \gamma_2 - \gamma_4 + \gamma_5 + \gamma_{10} - \gamma_{11} - \gamma_{12} + \gamma_{13} - 2\gamma_{14} + 2\gamma_{15}$$

The condition in Eq. (6.11.3), when imposed on Eq. (6.11.11), require that

$$\lambda_k = \alpha_0 n_{,k} + \alpha_1 e_{krs}\nu_{s,r}\ , \quad \alpha_2 = \alpha_3 = \alpha_4 = 0$$

Finally, upon using the relation $m_{kl} = -e_{lrs}\lambda_{k[rs]}$, the first stress moment, in Eq. (6.11.3), is given by

$$\lambda_k = \alpha_0 n_{,k} + \alpha_1 e_{krs}\nu_{s,r} \tag{6.11.14}$$

$$m_{kl} = 4\beta_0 e_{klr} n_{,r} + 4\alpha\nu_{r,r}\delta_{kl} + 4\beta\nu_{k,l} + 4\gamma\nu_{l,k} \tag{6.11.15}$$

A substitution in Eq. (5.5.16), which represents the symmetric part of the equation governing the balance of first stress moments, from Eqs. (6.11.7), (6.11.9) and (6.11.3) gives

$$[\lambda_{k,r} + (\lambda_1 - \eta_1)n]\delta_{lm} + \rho[l_{(lm)} - \dot{\sigma}_{(lm)}] = 0 \tag{6.11.16}$$

Use of Eq. (6.11.5) reduces this equation to

$$[\lambda_{r,r} + (\lambda_1 - \eta_1)n - \frac{1}{2}\rho j(\dot{n} + n^2)]\delta_{kl}$$

$$+ \rho[l_{(kl)} + \frac{1}{2}j(\nu_{rr}\,\delta_{kl} - \nu_r\,\nu_r)] = 0 \qquad (6.11.17)$$

Thus, if the symmetric part, $l_{(kl)}$, of the first body moment, l_{kl}, is assumed to be of the form

$$l_{(kl)} = \frac{1}{2}j(\nu_k\nu_l - \nu_r\nu_r\delta_{kl}) \qquad (6.11.18)$$

then the symmetric part of the first stress moment balance is satisfied if

$$\lambda_{r,r} - (\eta_1 - \lambda_1)n - \frac{1}{2}\rho j(\dot{n} + n^2) = 0 \qquad (6.11.19)$$

Finally, upon substituting from Eq. (6.11.7) in Eq. (5.5.8), from Eqs. (6.11.6), (6.11.7) and (6.11.15) in Eq. (6.3.8), and from Eq. (6.11.14) in Eq. (6.11.17), the equations of motion are given by

$$\rho a_k = -\pi_{,k} + \lambda_1 n_{,k} + (\lambda + \mu - \kappa)(v_{r,r})_{,k} + (\mu + \kappa)v_{k,rr}$$

$$+ 2\kappa e_{krs}\nu_{s,r} + \rho f_k \qquad (6.11.20)$$

$$\rho\dot{\sigma}_k = \rho j(\dot{\nu}_k + 2n\nu_k) = 4(\alpha + \beta)(\nu_{r,r})_{,k} + 4\gamma\nu_{k,rr}$$

$$+ 2\kappa e_{krs}n_{s,r} - 4\kappa\nu_k + \rho l_k \qquad (6.11.21)$$

$$\frac{1}{2}\rho j(\dot{n} + n^2) = \alpha_0 n_{,rr} - (\eta_1 - \lambda_1)n \qquad (6.11.22)$$

to which must be added the equations

$$\frac{D}{Dt}(j) - 2nj = 0 \qquad (6.11.23)$$

$$\dot{\rho} + \rho v_{r,r} = 0 \qquad (6.11.24)$$

Equations (6.11.20) to (6.11.24) are then nine equations in the nine unknowns: n_k, ν_k, n, j and ρ. When $n = 0$, these equations reduce to those of micropolar fluids.

A further simplification of the equations may be achieved by neglecting the nonlinear terms in the expressions for $\dot{\sigma}_{(kl)}$ and $\dot{\sigma}_{[kl]}$, resulting in

$$\dot{\sigma}_{(kl)} = \frac{1}{2}j\dot{n}\delta_{kl} \quad , \quad \dot{\sigma}_{[kl]} = -\frac{1}{2}je_{klr}\dot{\nu}_r \qquad (6.11.25)$$

Then, the left-hand sides of Eqs. (6.11.20) and (6.11.21) become $\rho j \dot{\nu}_k$ and $\frac{1}{2}\rho j \dot{n}$, respectively.

For such a "linear theory," the field equations are then

$$\dot{\rho} + \rho v_{r,r} = 0$$

$$\frac{D}{Dt}(j) - 2nj = 0$$

$$\rho a_k = -\pi_{,k} + \lambda_1 n_{,k} + (\lambda + \mu - \kappa)(v_{r,r})_{,k} + (\mu + \kappa) v_{k,rr} + 2\kappa e_{krs} \nu_{s,r} + \rho f_k$$

$$\rho j \dot{\nu}_k = 4(\alpha + \beta)(\nu_{r,r})_{,k} + 4\gamma \nu_{k,rr} + 2\kappa e_{krs} v_{s,r} - 4\kappa \nu_k + \rho l_k$$

$$\frac{1}{2}\rho j \dot{n} = \alpha_0 n_{,rr} - (\eta_1 - \lambda_1) n$$

The stresses are given by Eq. (6.11.7) and the first stress moments by Eq. (6.11.3), where λ_k and m_{kl} are given, respectively, by Eqs. (6.11.14) and (6.11.15).

The results of this section are based on the work of Eringen [Ref. 4], where the Clausius-Duhem inequality has been shown to require that the viscosities and the gyroviscosities satisfy the inequalities

$$\left.\begin{array}{l} 3\lambda + 2\mu \geqslant 0\,, \ \mu \geqslant 0\,, \ \kappa \geqslant 0 \\ 3\alpha + 2\gamma \geqslant 0\,, \ -\gamma \leqslant \beta \leqslant \gamma\,, \ \nu \geqslant 0 \\ \alpha_0 \geqslant 0\,, \ \eta_1 - \lambda_1 \geqslant 0\,, \ (\eta - \lambda_1)(3\lambda + 2\mu) \geqslant \dfrac{\lambda^2}{4} \end{array}\right\} \tag{6.11.26}$$

6.12 Concluding Remarks

This chapter has studied micropolar fluids and micropolar fluids with stretch, both of which are special types of micro fluids. Micropolar fluids are a subclass of micropolar fluids with stretch. Such fluids have microstructure and can support couple stresses. Stokesian fluids are a subclass of such fluids.

These theories are expected to be useful for describing the flow of fluids containing particulate structures and liquid crystals. These theories could also be used for modeling turbulence.

The theory of couple stresses in fluids in the absence of microstructure, described in Chapter 3, is a special case of the theory of micropolar fluids with the microrotation vector constrained to be equal to the vorticity. However, Ericksen's theory of anisotropic fluids, given in Chapter 4, is not a subclass of micropolar fluids.

6.13 References

The material in this chapter is mainly based on the first four references given below. References [5] and [6] pertain to Section (6.9).

A review of the relationship of the theory of micropolar fluids to other theories of fluids with microstructure, and some applications, are given in Refs. [7] and [8]. Two alternative approaches to polar theories, which have not been considered in this book, are discussed in Refs. [9] and [10]. The most complete accounts of polar theories, to date, are given in Refs. [11] and [12].

The remaining references, arranged in chronological order, represent *selected* papers that are relevant to the theory and applications of micropolar fluids. A more complete chronological list is given in the Bibliography.

1. Eringen, A. C. (1965). Theory of Micropolar Continua, *Dev. Mech.* ("Proc. Ninth Midwest. Mech. Conf., Madison"), (T. C. Huang, and M. W. Johnson, Jr., Eds.), John Wiley, 23-40.
2. Eringen, A. C. (1966). Theory of Micropolar Fluids, *J. Math. Mech.* **16**, 1-18.
3. Eringen, A. C. (1966). Mechanics of Micromorphic Materials, *Proc. 11th Int. Cong. Appl. Mech.* (H. Görtler, Ed.), Springer Verlag, 131-138.
4. Eringen, A. C. (1969). Micropolar Fluids with Stretch, *Int. J. Eng. Sci.* 7, 115-127.
5. Ariman, T., Cakmak, A. S., and Hill, L. R. (1967). Flow of Micropolar Fluids Between Two Concentric Cylinders, *Phys. Fluids* **10**, 2546-2550.
6. Cowin, S. C., and Pennington, C. J. (1970). On the Steady Rotational Motion of Polar Fluids, *Rheol. Acta* **8**, 307-312.
7. Ariman, T., Turk, M. A., and Sylvester, N. D. (1973). Microcontinuum Fluid Mechanics — A Review, *Int. J. Eng. Sci.* **11**, 905-930.
8. Ariman, T., Turk, M. A., and Sylvester, N. D. (1974). Review Article: Applications of Microcontinuum Fluid Mechanics, *Int. J. Eng. Sci.* **12**, 273-293.
9. Bleustein, J. L., and Green, A. E. (1966). Dipolar Fluids, *Int. J. Eng. Sci.* **5**, 323-340.
10. Cowin, S. C. (1974). The Theory of Polar Fluids, "Advances in Ap-

plied Mechanics," (Chia-Shun Yih, Ed.), Vol. 14, Academic Press, New York, 279-347.

11. Eringen, A. C. (Editor), (1976). "Continuum Physics, Vol. IV: Polar and Non-Local Field Theories," Academic Press, New York.
12. Eringen, A. C. (1980). Theory of Anisotropic Micropolar Fluids, *Int. J. Eng. Sci.* **18**, 5-17.
13. Peddieson, J., Jr., and McNitt, P. R. (1968). Boundary-Layer Theory for a Micropolar Fluid, "Recent Advances in Engineering Science," (A. C. Eringen, Ed.), Vol. 5/1, Gordon and Breach, New York, 405-426.
14. Eringen, A. C. (1968). Mechanics of Micromorphic Continua, "Mechanics of Generalized Continua," (E. Kröner, Ed.), Springer-Verlag, Berlin, 18-35.
15. Eringen, A. C., and Chang, T. S. (1968). A Micropolar Description of Hydrodynamic Turbulence, "Recent Advances in Engineering Science," (A. C. Eringen, Ed.), Vol. 5/2, Gordon and Breach, New York, 1-8.
16. Grot, R. A. (1969). Thermodynamics of a Continuum with Microstructures, *Int. J. Eng. Sci.* **7**, 801-814.
17. Eringen, A. C. (1970). Balance Laws of Micromorphic Mechanics, *Int. J. Eng. Sci.* **8**, 819-828.
18. Eringen, A. C. (1970). Mechanics of Micropolar Continua, "Contributions to Mechanics," (D. Abir, Ed.), Pergamon Press, 23-40.
19. Liu, C. Y. (1970). On Turbulent Flow of Micropolar Fluids, *Int. J. Eng. Sci.* **8**, 457-466.
20. Rao, S. K. L. (1970). Stability of Micropolar Fluid Motions, *Int. J. Eng. Sci.* **8**, 753-762.
21. Tokuoka, T. (1970). Optical Properties of Polarizable Linear Micropolar Fluids, *Int. J. Eng. Sci.* **8**, 31-37.
22. Wilson, A. J. (1970). Boundary Layers in Micropolar Liquids, *Proc. Cambridge Philos. Soc.* **67**, 469-476.
23. Allen, S. J., and Kline, K. A. (1971). Lubrication Theory of Micropolar Fluids, *J. Appl. Mech.* **38**, 646-649.
24. Ariman, T. (1971). On the Analysis of Blood Flow, *J. Biomech.* **4**, 185-192.
25. Kafadar, C. B., and Eringen, A. C. (1971). Micropolar Media — I. The Classical Theory, *Int. J. Eng. Sci.* **9**, 271-305.
26. Kafadar, C. B., and Eringen, A. C. (1971). Micropolar Media — II.

The Relativistic Theory, *Int. J. Eng. Sci.* **9**, 307-329.

27. Rao, S. K. L. (1971). Existence of Periodic Solutions of the Equations of Incompressible Micropolar Fluid Flow, *Int. J. Eng. Sci.* **9**, 1143-1150.

28. Eringen, A. C. (1972). Theory of Thermomicrofluids, *J. Math. Anal. Appl.* **39**, 253-266.

29. Eringen, A. C. (1972). Continuum Foundations of Rheology — New Adventures, "Heat and Mass Transfer," (W. R. Schowalter, et al., Eds.), Vol. 5, Pergamon Press, 1-18.

30. Kirwan, A. D., Jr., and Newman, N. (1972). Time Dependent Channel Flow of a Micropolar Fluid, *Int. J. Eng. Sci.* **10**, 137-146.

31. Peddieson, J., Jr. (1972). An Application of the Micropolar Fluid Model to the Calculation of a Turbulent Shear Flow, *Int. J. Eng. Sci.* **10**, 23-32.

32. Eringen, A. C. (1973). On Non Local Microfluid Mechanics, *Int. J. Eng. Sci.* **11**, 291-306.

33. Turk, M. A., Sylvester, N. D., and Ariman, T. (1973). On Pulsatile Blood Flow, *Trans. Soc. Rheol.* **17**, 1-21.

34. Ariman, T., Turk, N. A., and Sylvester, N. D. (1974). On Steady Pulsalite Flow of Blood, *J. Appl. Mech.* **41**, 1-7.

35. Ariman, T., Turk, M. A., and Sylvester, N. D. (1974). On Time Dependent Blood Flow, *Lett. Appl. Eng. Sci.* **2**, 21-36.

36. Ahmadi, G. (1975). Optical Properties of Polarizable Linear Magneto-Micropolar Fluid, *Int. J. Eng. Sci.* **13**, 209-215.

37. Ahmadi, G. (1975). Turbulent Shear Flow of Micropolar Fluids, *Int. J. Eng. Sci.* **13**, 959-964.

38. Prakash, J., and Sinha, P. (1975). Lubrication Theory of Micropolar Fluids and Its Application to a Journal Bearing, *Int. J. Eng. Sci.* **13**, 217-232.

39. Ramkissoon, H., and Majumdar, S. R. (1975). Creeping Flow of a Micropolar Fluid Past a Sphere, *Lett. Appl. Eng. Sci.* **3**, 133-142.

40. Ahmadi, G. (1976). Stability of Micropolar Fluid Layer Heated from Below, *Int. J. Eng. Sci.* **14**, 81-89.

41. Avudainayagam, A. (1976). The Effective Viscosity of a Dilute Suspension of Micropolar Fluid Particles in a Viscous Fluid, *Int. J. Eng. Sci.* **14**, 703-712.

42. Datta, A. B., and Sastry, V. U. K. (1976). Thermal Instability of a Horizontal Layer of Micropolar Fluid Heated from Below, *Int. J. Eng. Sci.* **14**, 769-788.

43. Kirwan, A. D., Jr., and Chang, M. S. (1976). On the Micropolar Ekman Problem, *Int. J. Eng. Sci.* **14**, 685-692.

44. Kirwan, A. D., Jr., Newman, N., and Chang, M. S. (1976). On Microdeformable Fluids: A Special Case of Microfluids, *Int. J. Eng. Sci.* **14**, 673-684.

45. Prakash, J., and Sinha, P. (1976). Squeeze Film Theory for Micropolar Fluids, *J. Lubr. Technol.* **98**, 139-144.

46. Prakash, J., and Christensen, H. (1977). Microcontinuum Theory for the Elastohydrodynamic Inlet Zone, *J. Lubr. Technol.* **99**, 24-29.

47. Tözeren, A., and Skalak, R. (1977). Micropolar Fluids as Models for Suspensions of Rigid Spheres, *Int. J. Eng. Sci.* **15**, 511-523.

48. Kuemmerer, H. (1978). Stability of Laminar Flows of Micropolar Fluids Between Parallel Walls, *Phys. Fluids* **21**, 1688-1693.

49. Sava, V. A. (1978). Initial-Boundary-Value Problems in Theory of Micropolar Fluids, *Z. Angew. Math. Mech.* **58**, 511-518.

50. Rao, S. K. L., and Raju, K. V. (1979). Stability of Microstretch Fluid Motions, *Int. J. Eng. Sci.* **17**, 465-473.

51. Niefer, R., and Kaloni, P. N. (1980). On the Motion of a Micropolar Fluid Drop in a Viscous Fluid, *J. Eng. Math.* **14**, 107-116.

52. Migun, N. P. (1981). Experimental Method of Determining Parameters Characterizing the Microstructure of Micropolar Liquids, *J. Eng. Phys. (USSR)* **41**, 832-835.

53. Mizukami, A. (1981). Nonsteady Shear Flow of Micropolar Films, *Int. J. Eng. Sci.* **19**, 75-82.

54. Sinha, P., Singh, C., and Prasad, K. R. (1982). Microcontinuum Analysis of the Self Propulsion of the Spermatozoa in Cervical Canal, *Int. J. Eng. Sci.* **20**, 1037-1048.

55. Sinha, P., Singh, C., and Prasad, K. R. (1982). Lubrication of Human Joints – A Microcontinuum Approach, *Wear* **80**, 159-181.

56. Kolpashchikov, V. L., Migun, N. P., and Prokhorenko, P. P. (1983). Experimental Determination of Material Micropolar Fluid Constants, *Int. J. Eng. Sci.* **21**, 405-411.

57. Olmstead, W. E., and Majumdar, S. R. (1983). Fundamental Oseen Solution for the 2-Dimensional Flow of a Micropolar Fluid, *Int. J. Eng. Sci.* **21**, 423-430.

APPENDIX

Notation

The purpose of this appendix is to explain the notation of Cartesian tensors, mainly using the *index* notation, which has been used in the book.

Let x_i and $\overline{x}_i$ be the rectangular Cartesian coordinates of the same point in space, relative to the two different Cartesian frames Ox_i and $\overline{Ox}_i$ with origins O and $\overline{O}$, respectively. Then, in general, these coordinates are related by a transformation of the form

$$\overline{x}_i = Q_{ir}x_r + b_i \ , \quad Q_{ir}Q_{jr} = \delta_{ij} \tag{A-1}$$

where the summation convention has been used, in which a summation over the range 1, 2, 3 is implied over pairs of repeated indices. Thus, in the above equation, $Q_{ir}x_r$ stands for $\sum_{s=1}^{3} Q_{is}x_s$ and $Q_{ir}Q_{jr}$ for $\sum_{s=1}^{3} Q_{is}Q_{jr}$. δ_{ij} is the Kronecker delta which equals one when $i = j$ and is zero otherwise.

It is evident that b_i is the position vector of the origin O relative to the frame $\overline{Ox}_i$. Q_{ij} can be shown to be the direction cosine of the $\overline{Ox}_i$ axis relative to the Ox_j axis. If a matrix notation is used such that $\overline{\mathbf{x}}$, $\mathbf{x}$, $\mathbf{b}$ and $\mathbf{Q}$ are the matrices

$$\overline{\mathbf{x}} = \left[\overline{x}_i\right] = \begin{bmatrix}\overline{x}_1\\ \overline{x}_2\\ \overline{x}_3\end{bmatrix}, \quad \mathbf{x} = \left[x_i\right], \quad \mathbf{b} = \left[b_i\right] \ , \quad \mathbf{Q} = \left[Q_{ij}\right] \tag{A-2}$$

then Eq. (A-1) can be written as the matrix equation

$$\overline{\mathbf{x}} = \mathbf{Qx} + \mathbf{b}, \ \ \mathbf{QQ}^T = I \tag{A-3}$$

where $\mathbf{I}$ is the identity matrix. Thus $\mathbf{Q}$ is an orthogonal matrix so that its determinant is given by det$\mathbf{Q} = \pm\ 1$. Whereas det$\mathbf{Q} = +\ 1$ corresponds to a pure rotation of the coordinate axes, det$\mathbf{Q} = -\ 1$ corresponds to a combination of a rotation and a reflection of the axes. The index and matrix notations which have been used in Eqs. (A-1) and (A-3) respectively, will be used interchangeably.

If $\overline{\mathbf{v}} = \left[\overline{v}_i\right]$ and $\mathbf{v} = \left[v_i\right]$ are the components of the same vector in the two Cartesian frames $\overline{Ox}_i$ and Ox_i, respectively, then these components are related by

$$\bar{v}_i = Q_{ir} v_r \quad or \quad \bar{\mathbf{v}} = \mathbf{Q}\mathbf{v} \tag{A-4}$$

There are objects which are almost like vectors: Under the coordinate transformation given in Eq. (A-1), the components of a pseudovector $\mathbf{w} = \left[w_i \right]$ transform as

$$\bar{w}_i = (\det\mathbf{Q})\, Q_{ir} w_r \quad or \quad \bar{\mathbf{w}} = (\det\mathbf{Q})\mathbf{Q}\mathbf{w} \tag{A-5}$$

where det**Q** is the determinant of **Q**. The Gibbsian cross-product of two vectors is an example of a pseudovector.

More generally an object **A**, having 3^n components $a_{i_1 i_2 \cdots i_n}$, where each of the i's can have the values 1, 2 and 3, is said to be a Cartesian tensor of order n if its components $\bar{\mathbf{A}} = \left[\bar{a}_{i_1 i_2 \cdots i_n} \right]$ in the transformed coordinate frame are given by

$$\bar{a}_{i_1 i_2 \cdots i_n} = Q_{i_1 j_1} Q_{i_2 j_2} \cdots Q_{i_n j_n} a_{j_1 j_2 \cdots j_n} \tag{A-6}$$

Similarly **W** is said to be a pseudotensor of order n if

$$\bar{w}_{i_1 i_2 \cdots i_n} = (\det\mathbf{Q})\, Q_{i_1 j_1} Q_{i_2 j_2} \cdots Q_{i_n j_n} w_{j_1 j_2 \cdots j_n} \tag{A-7}$$

A tensor of order one is also called a vector. In matrix notation, vectors are denoted by lower case bold type. Higher order tensors are denoted by upper case bold type, the order being inferred from the context.

The inner product of two vectors **u** and **v**, which in Gibbsian vector analysis is written as $\mathbf{u}\cdot\mathbf{v}$, is the scalar

$$u_r v_r \quad or \quad \mathbf{u}^T \mathbf{v} \tag{A-8}$$

If **n** is a vector and **T** a second order tensor, then $\mathbf{t} = \mathbf{T}\mathbf{n}$ is the vector $t_i = t_{ir} n_r$.

The third order alternating pseudo tensor, e_{ijk}, is defined by

$$e_{ijk} = \begin{cases} +1 & \text{if } (i,j,k) \text{ is an even permutation of } (1,2,3) \\ -1 & \text{if } (i,j,k) \text{ is an odd permutation of } (1,2,3) \\ 0 & \text{if two or more of the indices } i,\ j,\ k \text{ are equal} \end{cases} \tag{A-9}$$

The Gibbsian cross-product $\mathbf{w} = \mathbf{u}\times\mathbf{v}$ of two vectors **u** and **v** is then the pseudovector

$$w_i = e_{ijk} u_j v_k \tag{A-10}$$

A second order tensor $\mathbf{A} = \left[a_{ij} \right]$ is symmetric (skew-symmetric) if $a_{ij} = a_{ji}$ $(a_{ij} = -a_{ji})$. If **T** is a second order tensor, then its symmetric and skew-symmetric parts $\mathbf{T}^S$ and $\mathbf{T}^A$, respectively, are defined by

$$t_{ij}^S = t_{(ij)} = \frac{1}{2}(t_{ij} + t_{ji})$$

$$t_{ij}^A = t_{[ij]} = \frac{1}{2}(t_{ij} - t_{ji}) \tag{A-11}$$

so that

$$t_{ij} = t_{ij}^S + t_{ij}^A = t_{(ij)} + t_{[ij]} \tag{A-12}$$

More generally, parentheses and brackets are used in the sense

$$a_{i\ldots(j\ldots k)\ldots l} = \frac{1}{2}\left[a_{i\ldots j\ldots k\ldots l} + a_{i\ldots k\ldots j\ldots l}\right]$$

$$a_{i\ldots[j\ldots k]\ldots l} = \frac{1}{2}\left[a_{i\ldots j\ldots k\ldots l} - a_{i\ldots k\ldots j\ldots l}\right] \tag{A-13}$$

With any second order skew-symmetric tensor $\mathbf{W} = \left[w_{ij}\right]$, $w_{ij} = -w_{ji}$, can be associated a pseudovector $\mathbf{w} = \left[w_i\right]$ such that

$$w_i = \frac{1}{2}\, e_{ijk} w_{jk} \quad , \quad w_{ij} = e_{ijk} w_k \tag{A-14}$$

In terms of matrices, $\mathbf{W}$ and $\mathbf{w}$ are related by

$$\mathbf{W} = \begin{bmatrix} 0 & w_{12} & w_{13} \\ -w_{12} & 0 & w_{23} \\ -w_{13} & -w_{23} & 0 \end{bmatrix} = \begin{bmatrix} 0 & w_3 & -w_2 \\ -w_3 & 0 & w_1 \\ w_2 & -w_3 & 0 \end{bmatrix} \tag{A-15}$$

Derivatives of tensors are denoted by the "comma notation"

$$v_{j,i} = \frac{\partial v_i}{\partial x_i} \quad , \quad v_{k,ij} = \frac{\partial^2 v_k}{\partial x_i \partial x_j} \quad , \quad a_{ij\ldots k,l} = \frac{\partial a_{ij\ldots k}}{\partial x_l}$$

and so on. The divergence and curl of a vector $\mathbf{v}$ are then given, respectively, by $v_{r,r}$ and $e_{ijk} v_{k,j}$.

If B is a region in space, bounded by the surface ∂B, then for any tensor $A = \left[a_{ij\ldots k}\right]$, the generalized form of Gauss' theorem is given by

$$\int_B a_{ij\ldots k,r}\, dV = \int_{\partial B} a_{ij\ldots k} n_r\, dS \tag{A-16}$$

where $\mathbf{n} = \left[n_r\right]$ is the unit normal to the surface element dS.

Bibliography

Given below is a chronological listing of the references that are pertinent to theories of fluids with microstructure.

1909

Cosseràt, E. et F. "Theorie des Corps Deformables," Hermann and Cie, Paris.

1922

Jeffery, G. B. The Motion of Ellipsoidal Particles Immersed in a Viscous Fluid, *Proc. R. Soc. London Series A* **102**, 161-179.

1933

Oseen, C. W. The Theory of Liquid Crystals, *Trans. Farady Soc.* **29**, 883-899.

1947

Oldroyd, J. G. A Rational Formulation of the Equations of Plastic Flow for a Bingham Solid, *Proc. Cambridge Philos. Soc.* **43**, 100-105.

1950

Oldroyd, J. G. On the Formulation of Rheological Equations of State, *Proc. R. Soc. London Series A* **200**, 523-541.

1952

Grad, M. Statistical Mechanics, Thermodynamic and Fluid Dynamics of Systems, *Commun. Pure Appl. Math.* **5**, 455-494.

1954

Truesdell, C. "The Kinematics of Vorticity," Indiana University Press, Bloomington, Indiana.

1955

Noll, W. On the Continuity of Solid and Fluid States, *J. Ration. Mech. Anal.* **4**, 3-81.

1957

Smith, G. F., and Rivlin, R. S. The Anisotropic Tensors, *Q. J. Appl. Math.* **15**, 308-314.

1958

1. Frank, F. C. On the Theory of Liquid Crystals, *Discuss. Faraday Soc.* **25**, 19-28.
2. Guenther, W. Zur Statik und Kinematik der Cosseratschen Kontinuums, *Abh. Braunschw. Wiss. Ges.* **10**, 195-213.
3. Noll, W. A Mathematical Theory of the Mechanical Behaviour of Continuous Media, *Arch. Ration. Mech. Anal.* **2**, 197-226.

1960

1. Ericksen, J. L. Transversely Isotropic Fluids, *Kolloid-Z.* **173**, 117-122.
2. Ericksen, J. L. Anisotropic Fluids, *Arch. Ration. Mech. Anal.* **4**, 231-237.
3. Ericksen, J. L. Theory of Anisotropic Fluids, *Trans. Soc. Rheol.* **4**, 29-39.
4. Ericksen, J. L. A Vorticity Effect in Anisotropic Fluids, *J. Polym. Sci.* **47**, 327-331.
5. Truesdell, C., and Toupin, R. A. The Classical Field Theories, "Handbuch der Physik," (S. Flügge, Ed.), Vol. III/1, Springer-Verlag, Berlin.

1961

1. Cowin, S. C., and Jaunzemis, W. Kinematics of Cosserat Continua, "Air Force Office of Scientific Research," Technical Report No. 5, 1-83.
2. Dahler, J. S., and Scriven, L. E. Angular Momentum of Continua, *Nature* **192**, 36-37.
3. Ericksen, J. L. Conservation Laws of Liquid Crystals, *Trans. Soc. Rheol.* **5**, 23-34.
4. Ericksen, J. L. Poiseuille Flow of Certain Anisotropic Fluids, *Arch. Ration. Mech. Anal.* **8**, 1-8.
5. Hand, G. L. A Theory of Dilute Suspensions, *Arch. Ration. Mech. Anal.* **7**, 81-86.

1962

1. Ericksen, J. L. Orientation Induced by Flow, *Trans. Soc. Rheol.* **6**, 275-291.
2. Ericksen, J. L. Kinematics of Macromolecules, *Arch. Ration. Mech. Anal.* **9**, 1-8.
3. Ericksen, J. L. Hydrostatic Theory of Liquid Crystals, *Arch. Ration. Mech. Anal.* **9**, 371-378.
4. Ericksen, J. L. Nilpotent Energies in Liquid Crystal Theory, *Arch. Ration. Mech. Anal.* **10**, 189-196.
5. Eringen, A. C. "Nonlinear Theory of Continuous Media," McGraw-Hill, New York.
6. Mindlin, R. D., and Tiersten, H. F. Effects of Couple Stresses in Linear Elasticity, *Arch. Ration. Mech. Anal.* **11**, 415-448.
7. Verma, P. D. S. Couette Flow of Certain Anisotropic Fluids, *Arch. Ration. Mech. Anal.* **10**, 101-107.

1963

1. Ericksen, J. L. Singular Surfaces in Anisotropic Fluids, *Int. J. Eng. Sci.* **1**, 157-161.
2. Mindlin, R. D. Influence of Couple Stresses in Linear Elasticity, *Exp. Mech.* **20**, 1-7.

1964

1. Condiff, D. W., and Dahler, J. S. Fluid Mechanical Aspects of Antisymmetric Stress, *Phys. Fluids* **7**, 842-854.
2. Eringen, A. C., and Suhubi, E. S. Nonlinear Theory of Simple Micro-Elastic Solids — I, *Int. J. Eng. Sci.* **2**, 189-203.
3. Eringen, A. C. Simple Micro-Fluids, *Int. J. Eng. Sci.* **2**, 205-217.
4. Green, A. E. A Continuum Theory of Anisotropic Fluids, *Proc. Cambridge Philos. Soc.* **60**, 123-128.
5. Green, A. E., and Rivlin, R. S. Simple Force and Stress Multipoles, *Arch. Ration. Mech. Anal.* **16**, 325-353.
6. Green, A. E., and Rivlin, R. S. Multipolar Continuum Mechanics, *Arch. Ration. Mech. Anal.* **17**, 113-147.
7. Leslie, F. M. The Stability of Certain Anisotropic Fluids, *Proc. Cambridge Philos. Soc.* **60**, 949-955.
8. Mindlin, R. D. Micro-Structures in Linear Elasticity, *Arch. Ration. Mech. Anal.* **16**, 51-78.

1965

1. Aero, E. L., Bulygin, A. N., and Kuvshinskii, E. V. Asymmetric Hydromechanics, *J. App. Math. Mech. (USSR)* **29**, 333-346.
2. Coleman, B. D. Simple Liquid Crystals, *Arch. Ration. Mech. Anal.* **20**, 41-58.
3. Eringen, A. C. Theory of Micropolar Continua, "Developments in Mechanics" *(Proc. 9th Midwest. Mech. Conf., Madison)* (T. C. Huang, and M. W. Johnson, Jr., Eds.), John Wiley, New York, 23-40.
4. Green, A. E. Micro Materials and Micro-Polar Continuum Mechanics, *Int. J. Eng. Sci.* **3**, 533-538.
5. Kaloni, P. N. On Certain Steady Flows of Anisotropic Fluids, *Int. J. Eng. Sci.* **3**, 515-532.
6. Kaloni, P. N. On the Flow of Certain Orientable Fluids Between Two Coaxial Cones, *J. Fluid Mech.* **23**, 511-520.
7. Truesdell, C., and Noll, W. The Nonlinear Field Theories of Mechanics, "Handbuch der Physik," (S. Flügge, Ed.), Vol. III/3, Springer-Verlag, Berlin.

1966

1. Allen, S. J., and DeSilva, C. N. A Theory of Transversely Isotropic Fluids, *J. Fluid Mech.* **24**, 801-821.
2. Coleman, B. D., Markovitz, H., and Noll, W. "Viscometric Flows of Non-Newtonian Fluids," Springer-Verlag, Berlin.
3. Denn, M. M., and Metzner, A. B. Elementary Flows of Anisotropic Fluids, *Trans. Soc. Rheol.* **10**, 215-227.
4. Ericksen, J. L. Inequalities in Liquid Crystal Theory, *Phys. Fluids* **9**, 1205-1207.
5. Ericksen, J. L. Some Magnetohydrodynamic Effects in Liquid Crystals, *Arch. Ration. Mech. Anal.* **23**, 266-275.
6. Eringen, A. C. Theory of Micropolar Fluids, *J. Math. Mech.* **16**, 1-18.
7. Eringen, A. C. Mechanics of Micromorphic Materials, "Proc. 11th Int. Cong. Appl. Mech.," (H. Görtler, Ed.), Springer-Verlag, 131-138.
8. Kaloni, P. N. Certain Periodic Flows of Anisotropic Fluids, *Phys. Fluids* **9**, 1316-1321.
9. Leslie, F. M. Some Constitutive Equations for Anisotropic Fluids, *Q. J. Mech. Appl. Math.* **19**, 357-370.

10. Stokes, V. K. Couple Stresses in Fluids, *Phys. Fluids* **9**, 1710-1715.

1967

1. Allen, S. J., DeSilva, C. N., and Kline, K. A. Theory of Simple Deformable Directed Fluids, *Phys. Fluids* **10**, 2551-2555.
2. Ariman, T., and Cakmak, A. S. Couple Stresses in Fluids, *Phys. Fluids* **10**, 2497-2499
3. Ariman, T., Cakmak, A. S., and Hill, L. R. Flow of Micropolar Fluids Between Two Concentric Cylinders, *Phys. Fluids* **10**, 2545-2550
4. Bleustein, J. L., and Green, A. E. Dipolar Fluids, *Int. J. Eng. Sci.* **5**, 323-340.
5. Ericksen, J. L. Continuum Theory of Liquid Crystals, *Appl. Mech. Rev.* **20**, 1029-1032.
6. Ericksen, J. L. Twisting of Liquid Crystals, *J. Fluid Mech.* **27**, 59-64.
7. Eringen, A. C. Micropolar Fluids with Stretch, *Int. J. Eng. Sci.* **7**, 115-127.
8. Eringen, A. C. Linear Theory of Micropolar Viscoelasticity, *Int. J. Eng. Sci.* **5**, 191-205.
9. Hill, R. N. The Slow Flow of a Dipolar Fluid Past a Sphere, *Int. J. Eng. Sci.* **5**, 957-967.
10. Kirwan, A. D. Theory of Turbulent Eddies, *Phys. Fluids Supplement* 584-586.
11. Laws, N. A Simple Dipolar Curve, *Int. J. Eng. Sci.* **5**, 653-662.

1968

1. Ahmadi, G., Koh, S. L., and Goldschmidt, U. M. A Theory of Nonsimple Microfluids, "Recent Advances in Engineering Science," (A. C. Eringen, Ed.), Vol. 5/2, Gordon and Breach, New York, 9-20.
2. Allen, S. J., and Kline, K. A. Rectilinear Shear Flow of Fluids with Interacting Deformable Substructure, *Z. Angew. Math. Phys.* **19**, 425-433.
3. Ariman, T. Micropolar and Dipolar Fluids, *Int. J. Eng. Sci.* **6**, 1-8.
4. Askar, A., Cakmak, A. S., and Ariman, T. Linear Theory of Hereditary Micropolar Materials, *Int. J. Eng. Sci.* **6**, 283-294.
5. Bhatnagar, K. S., and Rajogopalan, R. Secondary Flows Introduced by Rotation of a Sphere or Coaxial Cones in Micropolar Fluids, *Rheol. Acta* **7**, 209-217.

6. Cowin, S. C. The Characteristic Length of a Polar Fluid, "Mechanics of Generalized Continua," (E. Kröner, Ed.), Springer-Verlag, Berlin, 90-94.

7. Cowin, S. C. Polar Fluids, *Phys. Fluids* **11**, 1919-1927.

8. Dafermos, C. M. Stability of Orientation Patterns of Liquid Crystals Subject to Magnetic Fields, *SIAM J. Appl. Math.* **16**, 1305-1318.

9. DeSilva, C. N., and Kline, K. A. Nonlinear Constitutive Equations for Directed Viscoelastic Materials with Memory, *Z. Angew. Math. Phys.* **19**, 128-139.

10. Ericksen, J. L. Twist Waves in Liquid Crystals, *Q. J. Mech. Appl. Math.* **21**, 463-465.

11. Ericksen, J. L. Twisting of Partially Oriented Liquid Crystals, *Q. Appl. Math.* **25**, 474-479.

12. Eringen, A. C. Mechanics of Micromorphic Continua, "Mechanics of Generalized Continua," (E. Kröner, Ed.), Springer-Verlag, Berlin, 18-35.

13. Eringen, A. C., and Chang, T. S. A Micropolar Description of Hydrodynamic Turbulence, "Recent Advances in Engineering Science," (A. C. Eringen, Ed.), Vol. 5/2, Gordon and Breach, New York, 1-8.

14. Kirwan, A. D., Jr. Constitutive Equations for a Fluid Containing Non-Rigid Structures, *Phys. Fluids* **11**, 1440-1446.

15. Kline, K. A., and Allen, S. J. On Continuum Theories of Suspensions of Deformable Particles, *Z. Angew. Math. Phys.* **19**, 898-905.

16. Kline, K. A., and Allen, S. J. Heat Conduction in Fluids with Substructure, *Z. Angew. Math. Mech.* **48**, 435-443.

17. Kline, K. A., Allen, S. J., and DeSilva, C. N. A Continuum Approach to Blood Flow, *Biorheology* **5**, 111-118.

18. Leslie, F. M. Some Constitutive Equations for Liquid Crystals, *Arch. Ration. Mech. Anal.* **28**, 265-283.

19. Leslie, F. M. Some Thermal Effects in Cholesteric Liquid Crystals, *Proc. R. Soc. London Series A* **307**, 359-372.

20. Nguen Van D'Ep Equations of Fluid Boundary Layer with Couple Stresses, *J. Appl. Math. Mech. (USSR)* **32**, 777-783.

21. Peddieson, J., Jr., and McNitt, P. R. Boundary-Layer Theory for a Micropolar Fluid, "Recent Advances in Engineering Science," (A. C. Eringen, Ed.), Vol. 5/1, Gordon and Breach, New York, 405-426.

22. Rajagopalan, R. Some Problems in Micropolar Fluids, *J. Indian Inst. Sci.* **50**, 57-77.

23. Stokes, V. K. Effects of Couple Stresses in Fluids on Hydromagnetic Channel Flows, *Phys. Fluids* **11**, 1131-1133.

24. Varma, P. D. S., and Sehgal, M. M. Couette Flow of Micropolar Fluids, *Int. J. Eng. Sci.* **6**, 233-238.

1969

1. Allen, S. J., and Kline, K. A. A Theory of Mixtures with Microstructures, *Z. Angew. Math. Phys.* **20**, 145-155.

2. Erdogan, M. E., and Gürgöze, I. T. Generalized Couette Flow of a Dipolar Fluid Between Two Parallel Plates, *Z. Angew. Math. Phys.* **20**, 785-790.

3. Ericksen, J. L. A Boundary Layer Effect in Viscometry of Liquid Crystals, *Trans. Soc. Rheol.* **13**, 9-15.

4. Ericksen, J. L. Continuum Theory of Liquid Crystals of the Nematic Type, *Mol. Cryst. Liq. Cryst.* **7**, 153-164.

5. Ericksen, J. L. Twisting of Liquid Crystals by Magnetic Fields, *Z. Angew. Math. Phys.* **20**, 383-388.

6. Eringen, A. C. Micropolar Fluids with Stretch, *Int. J. Eng. Sci.* **7**, 115-127.

7. Grot, R. A. Thermodynamics of a Continuum with Microstructures, *Int. J. Eng. Sci.* **7**, 801-814.

8. Hudimoto, B. Two Dimensional Shear Flows of Linear Micropolar Fluids, *Int. J. Eng. Sci.* **7**, 515-522.

9. Kaloni, P. N., and DeSilva, C. N. Oriented Fluids and the Rheology of Suspensions, *Phys. Fluids* **12**, 994-999.

10. Kaloni, P. N., and DeSilva, C. N. Elongational Flow of a Dilute Suspension of Viscoelastic Spheres, *Phys. Fluids* **12**, 2437-2438.

11. Kaloni, P. N., and DeSilva, C. N. On the Rheology of Dilute Suspensions of Rigid Particles, *Trans. Soc. Rheol.* **14**, 319-334.

12. Kirwan, A. D., and Newman, N. Simple Flow of a Fluid Containing Deformable Structures, *Int. J. Eng. Sci.* **7**, 1067-1078.

13. Kline, K. A., and Allen, S. J. Nonsteady Flows of Fluids with Microstructure, *Phys. Fluids* **13**, 263-270.

14. Kline, K. A., and Allen, S. J. The Relationship of Pressure Gradient to Blood Velocity Based on a Continuum Theory of Blood, *J. Biomech.* **2**, 313-318.

15. Leslie, F. M. Continuum Theory of Cholesteric Liquid Crystals, *Mol. Cryst. Liq. Cryst.* **7**, 407-420.

16. Pennington, C. J., and Cowin, S. C. Couette Flow of Polar Fluid, *Trans. Soc. Rheol.* **13**, 387-403.

17. Rao, S. K. L., Ramacharyulu, S. C. P., and Rao, P. B. Slow Steady Rotation of a Sphere in a Micropolar Fluid, *Int. J. Eng. Sci.* **7**, 905-916.

18. Stokes, V. K. On Some Effects of Couple Stresses in Fluids on Heat Transfer, *J. Heat Transfer* **91**, 182-184.

19. Valanis, K. C., and Sun, C. T. Poiseuille Flow of a Fluid with Couple Stresses with Applications to Blood Flow, *Biorheology* **6**, 85-97.

20. Wilson, A. J. Basic Flows of Micropolar Liquid, *Appl. Sci. Res.* **20**, 338-355.

1970

1. Allen, S. J., and Kline, K. A. Fluid Suspensions; Flow Near an Oscillating Plate, *Trans. Soc. Rheol.* **14**, 39-45.

2. Atkin, R. J. Poiseuille Flow of Liquid Crystals of the Nematic Type, *Arch. Ration. Mech. Analy.* **38**, 224-240.

3. Atkin, R. J., and Leslie, F. M. Couette Flow of Nematic Liquid Crystals, *Q. J. Mech. Appl. Math.* **23**, 53-524.

4. Bhatnagar, P. L., and Ravindran, R. A General Treatment of Two-Dimensional Plane Flows for a Micropolar Fluid, *Rheol. Acta* **9**, 218-222.

5. Cowin, S. C., and Pennington, C. J. On the Steady Rotational Motion of Polar Fluids, *Rheol. Acta* **9**, 307-312.

6. Currie, P. K. Couette Flow of a Nematic Liquid Crystal in the Presence of a Magnetic Field, *Arch. Ration. Mech. Anal.* **37**, 222-242.

7. Dafermos, C. M. Disinclinations in Liquid Crystals, *Q. J. Mech. Appl. Math.* **23**, S49-S64.

8. Erdogan, M. E. Polar Effects in Apparent Viscosity of a Suspension, *Rheol. Acta* **9**, 434-438.

9. Erdogan, M. E. A Continuum Model of Fluid Suspensions, *Bull. Tech. Univ. Istanbul* **23**, 27-36.

10. Eringen, A. C. Balance Laws of Micromorphic Mechanics, *Int. J. Eng. Sci.* **8**, 819-828.

11. Eringen, A. C. Mechanics of Micropolar Continua, "Contributions to Mechanics," (D. Abir, Ed.), Pergamon Press, 23-40.

12. Kaloni, P. N., and DeSilva, C. N. A Theory of Oriented Fluids, *Phys. Fluids* **13**, 1708-1716.

13. Kaloni, P. N., and DeSilva, C. N. On the Rheology of Dilute Suspensions of Rigid Particles, *Trans. Soc. Rheol.* **14**, 319-334.

14. Kirwan, A. D., Jr., and Chang, M. S. Cylindrical Flow of a Fluid Containing Deformable Structures, *Int. J. Eng. Sci.* **8**, 731-742.

15. Kline, S. J. The Reynolds — Orr Energy Equation with Applications to the Stability of Polar Fluid Motions, *Trans. Soc. Rheol.* **14**, 335-349.

16. Kline, K. A., and Allen, S. J. Nonsteady Flows of Fluids with Microstructure, *Phys. Fluids* **13**, 263-270.

17. Liu, C. Y. On Turbulent Flow of Micropolar Fluids, *Int. J. Eng. Sci.* **8**, 457-466.

18. Pennington, C. J., and Cowin, S. C. The Effective Viscosity of Polar Fluids, *Trans. Soc. Rheol.* **14**, 219-238.

19. Rao, S. K. L. Stability of Micropolar Fluid Motions, *Int. J. Eng. Sci.* **8**, 753-762.

20. Stojanovich, R. "Recent Developments in the Theory of Polar Continua," Springer-Verlag.

21. Tokuoka, T. Optical Properties of Polarizable Linear Micropolar Fluids, *Int. J. Eng. Sci.* **8**, 31-37.

22. Wilson, A. J. Boundary Layers in Micropolar Liquids, *Proc. Cambridge Philos. Soc.* **67**, 469-476.

1971

1. Ahmadi, G., Koh, S. L., and Goldschmidt, V. W. Mechanics of a Second-Order Microfluid, *Iran. J. Sci. Tech.* **1**, 233-257.

2. Allen, S. J., and Kline, K. A. A Thermodynamical Theory of Fluid Suspensions, *Phys. Fluids* **14**, 1863-1969.

3. Allen, S. J., and Kline, K. A. Lubrication Theory of Micropolar Fluids, *J. Appl. Mech.* **38**, 646-649.

4. Allen, S. J., Kline, K. A., and Ling, C. C. Kinematics of Deformable Suspensions, *Trans. Soc. Rheol.* **15**, 177-188.

5. Ariman, T. On the Analysis of Blood Flow, *J. Biomech.* **4**, 185-192.

6. Erdogan, M. E., and Kadioglu, N. The Viscosity of a Polar Fluid with Suspensions, *Rheol. Acta* **10**, 378-381.

7. Kafadar, C. B., and Eringen, A. C. Micropolar Media — I. The Classical Theory, *Int. J. Eng. Sci.* **9**, 271-305.

8. Kafadar, C. B., and Eringen, A. C. Micropolar Media — II. The Relativistic Theory, *Int. J. Eng. Sci.* **9**, 307-329.

9. Lee, J. D., and Eringen, A. C. Wave Propagation in Nematic Liquid Crystals, *J. Chem. Phys.* **54**, 5027-5034.

10. Lee, J. D., and Eringen, A. C. Alignment of Nematic Liquid Crystals, *J. Chem. Phys.* **55**, 4504-4508.

11. Lee, J. D., and Eringen, A. C. Boundary Effects of Orientation of Nematic Liquid Crystals, *J. Chem. Phys.* **55**, 4509-4512.

12. Rao, S. K. L. Existence of Periodic Solutions of the Equations of Incompressible Micropolar Fluid Flow, *Int. J. Eng. Sci.* **9**, 1143-1150.

13. Rao, S. K. L. Kinetic Energy of Incompressible Microstretch Fluid in a Domain Bounded by Rigid Walls, *Int. J. Eng. Sci.* **9**, 1151-1156.

14. Rao, S. K. L., and Rao, P. B. The Oscillations of a Sphere in Micropolar Fluid, *Int. J. Eng. Sci.* **9**, 651-672.

15. Soundalgekar, V. M. Effects of Couple Stresses in Fluids on Dispersion of a Solute in a Channel Flows, *Phys. Fluids* **14**, 19-20.

16. Stokes. V. K. Effects of Couple Stresses in Fluids on the Creeping Flow Past a Sphere, *Phys. Fluids* **14**, 1580-1582.

17. Twiss, R. J., and Eringen, A. C. Theory of Mixtures of Micromorphic Materials — I. Balance Laws, *Int. J. Eng. Sci.* **9**, 1019-1044.

1972

1. Chawla, S. S. Boundary Layer Growth of a Micropolar Fluid, *Int. J. Eng. Sci.* **10**, 981-987.

2. Cowin, S. C. On the Polar Fluid as a Model for Blood Flow in Tubes, *Biorheology* **9**, 23-25.

3. Cowin, S. C. Steady Flow of a Polar Fluid in Rotating Pipes, *Rheol. Acta* **11**, 330-332.

4. Erdogan, M. E. Dynamics of Polar Fluids, *Acta Mech.* **15**, 233-253.

5. Eringen, A. C. On Non Local Fluid Mechanics, *Int. J. Eng. Sci.* **10**, 561-575.

6. Eringen, A. C. Theory of Micromorphic Materials with Memory, *Int. J. Eng. Sci.* **10**, 623-641.

7. Eringen, A. C. Theory of Thermomicrofluids, *J. Math. Anal. Appl.* **39**, 253-266.

8. Eringen, A. C. Continuum Foundations of Rheology — New Adventures, "Heat and Mass Transfer," (W. R. Schowalter, et al., Eds.), Vol. 5, Pergamon Press, 1-18.

9. Gordon, R. J., and Schowalter, W. R. Anisotropic Fluid Theory: A

Different Approach to the Dumbell Theory of Dilute Polymer Solutions, *Trans. Soc. Rheol.* **16**, 79-97.

10. Jain, J. K., and Stokes, V. K. Effects of Couple Stresses on the Stability of Plane Poiseuille Flow, *Phys. Fluids* **15**, 977-980.

11. Kirwan, A. D., Jr., and Newman, N. Time Dependent Channel Flow of a Micropolar Fluid, *Int. J. Eng. Sci.* **10**, 137-146.

12. Kline, K. A., and Carmi, S. On the Stability of Motions of a Dilute Suspension of Rigid Spherical Particles: Universal Stability, *Bull. Acad. Pol. Sci. Ser. Sci. Tech.* **20**, 693-699.

13. Peddieson, J., Jr. An Application of the Micropolar Fluid Model to the Calculation of a Turbulent Shear Flow, *Int. J. Eng. Sci.* **10**, 23-32.

14. Rao, S. K. L., and Rao, P. B. Circular Cylinder Oscillating about a Mean Position in Incompressible Micropolar Fluid, *Int. J. Eng. Sci.* **10**, 185-191.

15. Ravindran, R., and Devi, R. G. A Study of Simple Shearing Flows in Polar Fluids, *Rheol. Acta* **11**, 99-105.

16. Shahinpoor, M., and Ahmadi, G. Stability of Cosserat Fluid Motions, *Arch. Ration. Mech. Anal.* **47**, 188-194.

17. Stokes, V. K. On the Analysis of Asymmetric Stress, *J. Appl. Mech.* **39**, 1133-1136.

1973

1. Allen, S. J., Kline, K. A., and Ling, C. C. Transient Shear Flow of Fluids with Deformable Microstructures, *Acta Mech.* **18**, 1-20.

2. Ariman, T., Turk, M. A., and Sylvester, N. D. Microcontinuum Fluid Mechanics — A Review, *Int. J. Eng. Sci.* **11**, 905-930.

3. Cowin, S. C., and Webb, G. R. Remarks on the Paper, 'Stability of Cosserat Fluid Motion', *Arch. Ration. Mech. Anal.* **52**, 118-120.

4. Currie, P. K. Static Solutions with Phase Change for a Nematic Liquid Crystal in a Pipe, *Trans. Soc. Rheol.* **17**, 197-208.

5. Ebert, F. Similarity Solution for the Boundary Layer Flow of a Polar Fluid, *Chem. Eng. J. (Lausanne)* **5**, 85-92.

6. Eringen, A. C. On Non Local Microfluid Mechanics, *Int. J. Eng. Sci.* **11**, 291-306.

7. Girija Devi, R. A Study of Shearing Flow in Polar Fluids, Part II — Flow Between Two Rotating Coaxial Cones, *Rheol. Acta* **12**, 53-57.

8. Hills, R. N. On the Stability of a Linear Dipolar Fluid, *Acta Mech.* **17**, 255-261.

9. Hills, R. N. On Uniqueness and Continuous Dependence for a Linear Micropolar Fluid, *Int. J. Eng. Sci.* **11**, 369-376.

10. Khader, M. S., and Vachon, R. I. Theoretical Effects of Solid Particles in Hydrostatic Bearing Lubricant, *J. Lubr. Technol.* **95**, 104-106.

11. Kline, K. A. Polymers as Structural Continua: Superposed Oscillatory Shear Flows, *Trans. Soc. Rheol.* **17**, 525-536.

12. Mani, R. Creeping Flow Induced by a Spinning Sphere Including Couple Stresses, *Acta Mech.* **18**, 81-88.

13. Shahinpoor, M., and Ahmadi, G. Decay of the Kinetic Energy of Compressible Micropolar Fluids, *Int. J. Eng. Sci.* **11**, 885-889.

14. Suyazov, V. M. Magnetohydrodynamics of Microstructural Medium: The Rheoelectric Effect, *Magnetohydrodynamics (USSR)* **9**, 486-491.

15. Turk, M. A., Sylvester, N. D., and Ariman, T. On Pulsatile Blood Flow, *Trans. Soc. Rheol.* **17**, 1-21.

1974

1. Ahmadi, G., and Shahinpoor, M. Universal Stability of Magneto-Micropolar Fluid Motions, *Int. J. Eng. Sci.* **12**, 657-663.

2. Ariman, T., Turk, M. A., and Sylvester, N. D. Review Article: Applications of Microcontinuum Fluid Mechanics, *Int. J. Eng. Sci.* **12**, 273-293.

3. Ariman, T., Turk, N. A., and Sylvester, N. D. On Steady Pulsalite Flow of Blood, *J. Appl. Mech.* **41**, 1-7.

4. Ariman, T., Turk, M. A., and Sylvester, N. D. On Time Dependent Blood Flow, *Lett. Appl. Eng. Sci.* **2**, 21-36.

5. Barratt, P. J. A Continuum Model for Disinclination Lines in Nematic Liquid Crystals, *Q. J. Mech. Appl. Math.* **27**, 505-522.

6. Barratt, P. J., and Jenkins, J. T. Interfacial Effects in the Static Theory of Nematic Liquid Crystals, *Q. J. Mech. Appl. Math.* **27**, 11-127.

7. Cowin, S. C. The Theory of Polar Fluids, "Advances in Applied Mechanics," (Chia-Shun Yih, Ed.), Vol. 14, Academic Press, New York, 279-347.

8. Ericksen, J. L. Liquid Crystals and Cosserat Surfaces, *Q. J. Mech. Appl. Math.* **27**, 213-219.

9. Levitskii, S. P., and Listrov, A. T. Stability of Flow of a Liquid Crystal Layer on an Inclined Plane, *J. Appl. Math. Mech. (USSR)* **38**, 978-987.

10. Soundalgekar, V. M., and Aranake, R. N. Effects of Couple Stresses on the Oscillatory Flow Past an Infinite Plate with Constant Suction, *Meccanica* **9**, 194-198.

1975

1. Ahmadi, G. Optical Properties of Polarizable Linear Magneto-Micropolar Fluid, *Int. J. Eng. Sci.* **13**, 209-215.

2. Ahmadi, G. Turbulent Shear Flow of Micropolar Fluids, *Int. J. Eng. Sci.* **13**, 959-964.

3. Balaram, M. Micropolar Squeeze Films, *J. Lubr. Technol.* **97**, 635-637.

4. Cook R. D., and Koh, S. L. Viscous Fluids with Microstructure and Microtemperature, *Proc. 12th Annual Meeting Soc. Eng. Sci.* 81-97.

5. Hartmann, C. Second Order Effect in Micropolar Flows with Parallel Streamlines, *J. Mec.* **14**, 679-704.

6. Kline, K. A. Fluid Microstructure Strain in Shear Flows, *Acta Mech.* **21**, 209-219.

7. Kline, K. A. Predictions from Polar Fluid Theory Which are Independent of Spin Boundary Conditions, *Trans. Soc. Rheol.* **19**, 139-145.

8. Maiti, G. Convective Heat Transfer in Micropolar Fluid Flow Through a Horizontal Parallel Plate Channel, *Z. Angew. Math. Mech.* **55**, 105-111.

9. Narasimhan, M. N. L., and Eringen, A. C. Thermo-Micropolar Theory of Heat Conducting Nematic Liquid Crystals, *Int. J. Eng. Sci.* **13**, 233-246.

10. Nath, G. Similar Solutions for the Incompressible Laminar Boundary Layer with Pressure Gradient in Micropolar Fluids, *Rheol. Acta* **14**, 850-857.

11. Prakash, J., and Sinha, P. Lubrication Theory of Micropolar Fluids and Its Application to a Journal Bearing, *Int. J. Eng. Sci.* **13**, 217-232.

12. Ramkissoon, H., and Majumdar, S. R. Creeping Flow of a Micropolar Fluid Past a Sphere, *Lett. Appl. Eng. Sci.* **3**, 133-142.

13. Riha, P. On the Theory of Heat-Conducting Micropolar Fluids with Microtemperatures, *Acta Mech.* **23**, 1-8.

14. Shukla, J. A., and Isa, M. Generalized Reynolds Equation for Micropolar Lubricants and Its Application to Optimum One-Dimensional Slider Bearings: Effects of Solid-Particle Additives in Solution, *J. Mech. Eng. Sci.* **17**, 280-284.

15. Soundalgekar, V. M. Effects of Couple Stresses in Fluids on the Dispersion of a Soluble Matter in a Channel Flow with Homogeneous and Heterogeneous Reactions, *Int. J. Heat Mass Transfer* **18**, 527-530.

1976

1. Ahmadi, G. Stability of Micropolar Fluid Layer Heated from Below, *Int. J. Eng. Sci.* **14**, 81-89.

2. Ahmadi, G. Self-Similar Solution of Incompressible Micropolar Boundary Layer Flow Over a Semi-Infinite Plate, *Int. J. Eng. Sci.* **14**, 639-646.

3. Ahmadi, G. Universal Stability of Thermo-Magneto-Micropolar Fluid Motions, *Int. J. Eng. Sci.* **14**, 853-859.

4. Ahmadi, G. Universal Stability of Thermo-Diffusive-Magneto-Micropolar Fluid Motions, *Acta Tech. CSAV.* **21**, 623-633.

5. Annapurna, N., and Ramanaiah, G. Effect of Couple Stresses on the Unsteady Drainage of a Micro-Polar Fluid on a Flat Surface, *Jpn. J. Appl. Phys.* **15**, 2441-2444.

6. Avudainayagam, A. The Effective Viscosity of a Dilute Suspension of Micropolar Fluid Particles in a Viscous Fluid, *Int. J. Eng. Sci.* **14**, 703-712.

7. Chaturani, P. Some Comments on Poiseuille Flow of a Liquid with Couple Stresses with Applications to Blood Flow, *Biorheology* **13**, 133-136.

8. Datta, A. B., and Sastry, V. U. K. Thermal Instability of a Horizontal Layer of Micropolar Fluid Heated from Below, *Int. J. Eng. Sci.* **14**, 769-788.

9. Debnath, L. On a Microcontinuum Model of Pulsatile Blood Flow, *Acta Mech.* **24**, 165-177.

10. Ericksen, J. L. On Equations of Motion for Liquid Crystals, *Q. J. Mech. Appl. Math.* **29**, 203-208.

11. Eringen, A. C. (Editor), "Continuum Physics, Vol. IV: Polar and Non-Local Field Theories," Academic Press, New York.

12. Guram, G. S., and Smith, A. C. Rectilinear Pipe Flow of a Micropolar Fluid, *Utilitas Math.* **9**, 147-160.

13. Kirwan, A. D., Jr., and Chang, M. S. On the Micropolar Ekman Problem, *Int. J. Eng. Sci.* **14**, 685-692.

14. Kirwan, A. D., Jr., Newman, N., and Chang, M. S. On Microdeformable Fluids: A Special Case of Microfluids, *Int. J. Eng. Sci.* **14**, 673-684.

15. Lanege, H. Die Existenz von Lösungen der Gleichungen, Welche die Strömunginkompressibler mikropolarer Flüssigkeiten beschreiben, *Z. Angew. Math. Mech.* **56**, 129-139.

16. Laskowski, L. K., and Wronski, S. On the Pulsatile Flow of Micropolar Fluid, *Arch. Mech.* **28**, 615-624.

17. Leslie, F. M. Analysis of a Flow Instability in Nematic Liquid Crystals, *J. Phys. D* **9**, 925-937.

18. Prakash, J., and Sinha, P. Squeeze Film Theory for Micropolar Fluids, *J. Lubr. Technol.* **98**, 139-144.

19. Prakash, J., and Sinha, P. Cyclic Squeeze Films in Micropolar Fluid Lubricated Journal Bearings, *J. Lubr. Technol.* **98**, 412-417.

20. Ramkissoon, H., and Majumdar, S. R. Representation and Fundamental Singular Solutions in Micropolar Fluids, *Z. Angew. Math. Mech.* **56**, 197-203.

21. Sandberg, T. K., and Kline, K. A. Structured Fluid Theory, *Acta Mech.* **24**, 141-155.

22. Shahinpoor, M. On the Stress Tensor in Nematic Liquid Crystals, *Rheol. Acta* **15**, 99-103.

23. Suyazov, V. M. Hydrodynamics of a Fluid with a Deformable Microstructure, *Sov. Appl. Mech.* **11**, 1108-1115.

24. Tandon, P. N., and Pokhariyal, S. C. Velocity Distribution and Heat Transfer in a Micro-Polar Fluid Flowing in a Tapered Channel, *Indian J. Technol.* **14**, 596-599.

25. Vidyanidi, V., and Murty, M. S. The Dispersion of a Chemical Reacting Solute in a Micropolar Fluid, *Int. J. Eng. Sci.* **14**, 1127-1134.

1977

1. Batra, R. C. Decay of the Kinetic and the Thermal Energy of Incompressible Micropolar Fluids, *J. Mec.* **16**, 217-225.

2. Chaturani, P., and Upadhya, V. S. Gravity Flow of a Fluid with Couple Stress Along an Inclined Plane, *Biorheology* **14**, 237-246.

3. Datta, A. B., and Sastry, V. U. K. Flow of a Micropolar Liquid Between Two Slowly Rotating Disks, *Z. Angew. Math. Mech.* **58**, 113-115.

4. Dmitriev, N. M. Anisotropy Effects in Polymer Solution Flow, *Moscow Univ. Mech. Bull.* **32**, 66-70.

5. Kline, K. A., and Sandberg, T. K. A Polar Fluid Estimate of Relative Force, *Acta Mech.* **26**, 201-222.

6. Kline, K. A. A Spin-Vorticity Relation for Unidirectional Polar Flows of Micropolar Fluids, *Int. J. Eng. Sci.* **15**, 131-134.

7. Prakash, J., and Christensen, H. Microcontinuum Theory for the Elastohydrodynamic Inlet Zone, *J. Lubr. Technol.* **99**, 24-29.

8. Prakash, J., and Sinha, P. Theoretical Effects of Solid Particles on the Lubrication of Journal Bearings Considering Cavitation, *Wear* **41**, 233-249.

9. Radhakrishnama Charya, G. Flow of Micropolar Fluid Through a Constricted Channel, *Int. J. Eng. Sci.* **15**, 719-725.

10. Ramanaiah, G., and Dubey, J. N. Optimum Slider Profile of a Slider Bearing Lubricated with a Micropolar Fluid, *Wear* **42**, 1-7.

11. Ramkissoon, H. Slow Steady Rotation of an Axially Symmetric Body in a Micropolar Fluid, *Appl. Sci. Res.* **33**, 243-257.

12. Ramkissoon, H. Singular Solutions in Microcontinuum Fluid Mechanics, *Arch. Mech.* **29**, 115-123.

13. Ramkissoon, H., and Majumdar, S. R. Unsteady Flow of a Micro-Polar Fluid Between Two Concentric Circular Cylinders, *Can. J. Chem. Eng.* **55**, 408-413.

14. Riha, P. Heat Transfer in Fluids with Inner Structure, *Vodohospod. Cas.* **25**, 380-389.

15. Shahinpoor, M., and Ahmadi, G. Stability of Cosserat Fluid Motions — II on the N-th-Order Cosserat Fluid, *Acta Mech.* **28**, 153-164.

16. Sinha, P. Effects of Rigid Particles in the Lubrication of Rolling Contact Bearing Considering Cavitation, *Wear* **44**, 295-309.

17. Sinha, P. Dynamically Loaded Micropolar Fluid Lubricated Journal Bearings with Special Reference to Squeeze Films Under Fluctuating Loads, *Wear* **45**, 279-292.

18. Smith, A. C., and Guram, G. S. Geometrical Aspects of Steady Ekman Flow of a Micro-Polar Fluid, *Lett. Appl. Eng. Sci.* **5**, 215-227.

19. Soundalgekar, V. M., and Aranake, R. N. Effects of Couple Stresses in MHD Channel Flow, *Nucl. Eng. Des.* **44**, 301-308.

20. Tözeren, A., and Skalak, R. Micropolar Fluids as Models for Suspensions of Rigid Spheres, *Int. J. Eng. Sci.* **15**, 511-523.

1978

1. Ahmadi, G. Heat Conducting Microfluids, *Indian J. Technol.* **16**, 15-17.

2. Ahmadi, G. On the Theory of Extensible Nematic Liquid Crystals, *Mol. Cryst. Liq. Cryst.* **47**, 209-223.

3. Batra, R. C. Decay of Kinetic and Thermal Energy of Compressible Micropolar Fluids, *Acta Mech.* **29**, 47-53.

4. Chaturani, P. Viscosity of Poiseuille Flow of a Couple Stress Fluid with Applications to Blood Blow, *Biorheology* **15**, 119-128.

5. Chaturani, P., and Upadhya, V. S. Pulsatile Flow of a Couple Stress Fluid Through Circular Tubes with Applications to Blood Flow, *Biorheology* **15**, 193-201.

6. Dubois-Violette, E., and Manneville, P. Stability of Couette Flow in Nematic Liquid Crystals, *J. Fluid Mech.* **89**, 273-303.

7. Duffy, B. R. Flow of a Liquid with an Anisotropic Viscosity Tensor, *J. Non-Newtonian Fluid Mech.* **4**, 177-193.

8. Erdogan, M. E. On a Non-Newtonian Fluid, *Acta Mech.* **30**, 79-88.

9. Indrasena, A. Some Kinematic Properties of Steady Micropolar Fluid Flows, *Jpn. J. Appl. Phys.* **17**, 2147-2150.

10. Kuemmerer, H. Stability of Laminar Flows of Micropolar Fluids Between Parallel Walls, *Phys. Fluids* **21**, 1688-1693.

11. Mathur, M. N., Ojha, S. K., and Ramachandran, P. S. Thermal Boundary Layer of a Micropolar Fluid on a Circular Cylinder, *Int. J. Heat Mass Transfer* **21**, 923-933.

12. Narasimhan, M. N. L., and Okada, K. Asymptotic Boundary Layer on a Circular Cylinder in Axisymmetric Micropolar Fluid Flow, *Int. J. Eng. Sci.* **16**, 1045-1059.

13. Ojha, S. K., Banerjee, A. K., and Mathur, M. N. Micropolar Fluid Jet Impingement on a Curved Surface, *Int. J. Eng. Sci.* **16**, 743-764.

14. Ramanaiah, G., and Sarkar, P. Squeeze Films and Thrust Bearings Lubricated by Fluids with Couple Stress, *Wear* **48**, 309-316.

15. Ramanaiah, G., and Sarkar, P. Optimum Load Capability of a Slider Bearing Lubricated by a Fluid with Couple Stress, *Wear* **49**, 61-66.

16. Rammkissoon, H. Drag in Couple Stress Fluids, *Z. Angew. Math. Phys.* **29**, 341-346.

17. Sava, V. A. Initial-Boundary-Value Problems in Theory of Micropolar Fluids, *Z. Angew. Math. Mech.* **58**, 511-518.

18. Zaheeruddin, Kh., and Isa, M. Micropolar Fluid Lubrication of One-Dimensional Journal Bearings, *Wear* **50**, 211-220.

19. Zaheeruddin, Kh., and Isa, M. Characteristics of a Micropolar Lubricant in a Squeeze Film Porous Spherical Bearing, *Wear* **51**, 1-10.

1979

1. Ahmadi, G. Special Theories of Heat Conducting Microfluids, *Bull. Acad. Pol. Sci. Ser. Sci. Tech.* **27**, 69-80.
2. Bader, P., Fodemski, T., and Rutkowski, J. Evaluation of Micropolar Properties of Some Liquids from Two Particular Flows, *Arch. Budowy. Masz.* **26**, 331-347.
3. Blanschong, G. P., and Hartmann, C. Study of Motions of Micropolar Fluids Produced by Orientable Particles, *J. Mec.* **18**, 713-744.
4. Currie, P. K. Apparent Viscosity During Viscometric Flow of Nematic Liquid Crystals, *J. Phys. (Paris)* **40**, 501-505.
5. Currie, P. K., and MacSithigh, G. P. Stability and Dissipation of Solutions for Shearing Flow of Nematic Liquid Crystals, *Q. J. Mech. Appl. Math.* **32**, 499-511.
6. Guram, G. S., and Anwar, M. Steady Flow of a Micropolar Fluid Due to a Rotating Disc, *J. Eng. Math.* **13**, 223-234.
7. Kamel, M. T., Kaloni, P. N., and Tory, E. M. Two-Dimensional Internal Flows of Polar Fluids, *J. Rheol.* **23**, 141-150.
8. Kolpashchikov, V. L., Migun, N. P., Prokhorenko, P. P., and Lis, V. I. Heat Exchange in a Microstructural Fluid Under Boundary Conditions of the Third Kind, *J. Eng. Phys. (USSR)* **37**, 785-789.
9. Kolpashchikov, V. L., Migun, N. P., Prokhorenko, P. P., and Lis, V. I. Unsteady Heat Transfer in a Micropolar Fluid Flowing in a Plane Channel, *J. Eng. Phys. (USSR)* **37**, 1405-1409.
10. Korzhov, E. N., and Listrov, A. T. Boundary Layer of a Microstructural Fluid on a Plate, *J. Eng. Phys. (USSR)* **36**, 88-92.
11. Manneville, P. Theoretical Analysis of Poiseuille Flow Instabilities in Nematics, *J. Phys. (Paris)* **40**, 713-724.
12. Ojha, S. K., Ramachandran, P. S., and Mathur, M. N. Thermal Boundary Layer of a Micropolar Fluid Jet Impinging Normally on a Flat Plate, *Acta Mech.* **33**, 281-294.
13. Ramachandran, P. S., Mathur, M. N., and Ojha, S. K. Heat Transfer in Boundary Layer Flow of a Micropolar Fluid Past a Curved Surface with Suction and Injection, *Int. J. Eng. Sci.* **17**, 625-639.
14. Ramanaiah, G., and Sarkar, P. Slider Bearings Lubricated by Fluids with Couple Stress, *Wear* **52**, 27-36.
15. Rao, S. K. L., and Raju, K. V. Stability of Microstretch Fluid Motions, *Int. J. Eng. Sci.* **17**, 465-473.
16. Riha, P. Theory of Mixtures for Microfluids, *Czech. Tech. CSAV* **24**, 389-396.

17. Sastry, V. U. K., and Rao, V. R. M. Micropolar Fluid Flow Due to an Oscillating Plane Subject to Rotation, *Acta Mech.* **33**, 45-53.

18. Tanahashi, T., Ardo, T., and Sawada, T. Flow of Polar Fluids in a Circular Pipe (1st Report, Oscillatory Flow Without Spin on the Pipe Wall), *Trans. JSME* **45**, 20-27.

19. Verma, P. D. S., Agrawal, V. K., and Bhatt, S. Porous Inclined Slider Bearing Lubricated with Micropolar Fluid, *Wear* **53**, 101-106.

20. Walicki, E., and Walicki, A. Throughflow of Micropolar Fluid in a Slot Between Surfaces of Revolution, *Rev. Roum. Sci. Tech. — Ser. Mech. Appl.* **24**, 581-588.

1980

1. Agrawal, V. K., and Bhatt, S. B. Porous Pivoted Slider Bearings Lubricated with a Micropolar Fluid, *Wear* **61**, 1-8.

2. Akay, G. Mass Transfer in Time Dependent Blood Flow, *Proc. 8th Int. Cong. Rheol. (Naples)* Vol. 3, Plenum Press, New York, 489-494.

3. Anwar, M., and Guram, G. S. Numerical Solution of a Micropolar Flow Between a Rotating and a Stationary Disc, *Comput. Math. Appl.* **6**, 235-245.

4. Bernardy, T. The Filtration of Micropolar Liquid Through the Porous Medium, *Vodohospod. Cas.* **28**, 319-326.

5. Duffy, B. R. Flow of a Liquid with an Anisotropic Viscosity Tensor: Inertial Effects, *J. Non-Newtonian Fluid Mech.* **7**, 107-120.

6. Duffy, B. R. Flow of a Liquid with an Anisotropic Viscosity Tensor: Some Axisymmetric Flows, *J. Non-Newtonian Fluid Mech.* **7**, 359-367.

7. Eringen, A. C. Theory of Anisotropic Micropolar Fluids, *Int. J. Eng. Sci.* **18**, 5-17.

8. Gorla, R. S. R. Thermal Boundary Layer of a Micropolar Fluid at a Stagnation Point, *Int. J. Eng. Sci.* **18**, 611-617.

9. Guram, G. S., and Smith, A. C. Stagnation Flows of Micropolar Fluids with Strong and Weak Interactions, *Comput. Math. Appl.* **6**, 213-233.

10. Isa, M., and Zaheeruddin, Kh. One-Dimensional Porous Journal Bearings Lubricated with Micropolar Fluid, *Wear* **63**, 257-270.

11. Lakshmana Rao, S. K., and Venkatapathi Raju, K. Variational Algorithm for the Stability of the Flow of Micropolar Fluids with Stretch, *Int. J. Eng. Sci.* **18**, 1411-1419.

12. Migun, N. P. Use of the WKB Method to Investigate Convective Heat Transfer in a Micropolar Fluid, *J. Eng. Phys.* **39**, 1146-1150.

13. Mills, P., Rubi, J. M., and Quemada, D. Suspensions Flow Described by Means of a Micropolar Fluid Theory and Apparent Viscosity of Aggregable Particles Suspension in a Couette Flow, *Proc. 8th Int. Cong. Rheol. (Naples)*, Vol. 2, Plenum Press, New York, 639-643.

14. Niefer, R., and Kaloni, P. N. On the Motion of a Micròpolar Fluid Drop in a Viscous Fluid, *J. Eng. Math.* **14**, 107-116.

15. Prakash, J., and Christensen, K. T. H. Micropolarity-Roughness Interaction in Hydrodynamic Lubrication, *J. Lubr. Technol.* **102**, 368-373.

16. Ramkissoon, H. On the Uniqueness and Existence of Stokes Flows in Micropolar Fluid Theory, *Acta Mech.* **35**, 259-270.

17. Ramkissoon, H. A Paradox in Microcontinuum Fluid Mechanics, *Q. Appl. Math.* **38**, 357-362.

18. Ramkissoon, H., and Devanathan, R. Drag Invariance in Microcontinuum Fluid Mechanics, *Acta Mech.* **37**, 145-151.

19. Ramkissoon, H., and Devanathan, R. On a Linear Steady-State System of Equations in Microcontinuum Fluid Mechanics, *Int. J. Eng. Sci.* **18**, 947-956.

20. Ramkissoon, H., and Majumdar, S. R. Potentials and Green's Functions in Micropolar Fluid Theory, *Z. Angew. Math. Mech.* **60**, 249-255.

21. Rao, G. T., and Moizuddin, M. Steady Flow of Micropolar Incompressible Fluid Between Two Parallel Porous Plates, *Def. Sci. J.* **30**, 105-112.

22. Rubi, J. M., and Casas-Vazquez, J. Thermodynamical Aspects of Micropolar Fluids — A Non-Linear Approach, *J. Non-Equilib. Thermodyn.* **5**, 155-164.

23. Rubi, J. M., Mills, P., Quemada, D., and Casas-Vazquez, J. On Boundary Conditions for Micropolar Fluids, *Proc. 8th Int. Cong. Rheol. (Naples)*, Vol. 2, Plenum Press, New York, 645-650.

24. Soundalgekar, V. M., and Takhar, S. Heat Transfer in Wedge Flow of a Micropolar Fluid, *Proc. 8th Int. Cong. Rheol. (Naples)*, Vol. 1, Plenum Press, New York, 321-325.

25. Stokes, V. K. On the Proper Vectors of Real Third Order Matrices, *J. Math. Phys. (N. Y.)* **21**, 2842-2843.

26. Zaheeruddin, Kh. Squeeze-Film Narrow Porous Journal Bearings Lubricated with a Micropolar Liquid, *Wear* **64**, 163-174.

1981

1. Ahmadi, G. On Statistical Theories of Turbulent Flow of Micropolar Fluids, *Acta Mech.* **39**, 127-138.

2. Ahmadi, G. Micropolar Model of Turbulence and Free Shear Flows, *Strojnicky Cas.* **32**, 409-424.

3. Ahmadi, G. Micropolar Fluid Model of Incompressible Turbulent Flow, *Indian J. Technol.* **19**, 425-428.

4. Barratt, P. J. Calculation of Shear Rate Thresholds for Homogeneous and Roll Type Instabilities in Nematics Subjected to Simple Shear Flow, *J. Phys. D* **14**, 1831-1846.

5. Demiray, H. Constitutive Equation for the Nematic Liquid Crystals, *J. Tech. Phys.* **22**, 217-232.

6. Duffy, B. R. Flow of a Liquid with an Anisotropic Viscosity Tensor: A Skew Initial Orientation, *J. Non-Newtonian Fluid Mech.* **8**, 213-222.

7. Duffy, B. R. Flow of a Liquid with an Anisotropic Viscosity Tensor: Elongational Flow, *J. Non-Newtonian Fluid Mech.* **9**, 1-12.

8. Guram, G. S., and Anwar, M. Micropolar Flow Due to a Rotating Disc with Suction and Injection, *Z. Angew. Math. Mech.* **61**, 589-595.

9. Isa, M., and Zaheeruddin, Kh. Characteristics of a Squeeze Film Porous Bearings with Micropolar Lubricant, *Trans. Can. Soc. Mech. Eng.* **6**, 146-150.

10. Lakahmana Rao, S. K., and Iyengar, T. K. V. Slow Steady Rotation of a Spheroid in an Incompressible Micropolar Fluid, *Int. J. Eng. Sci.* **19**, 655-687.

11. Leslie, F. M. Viscometry of Nematic Liquid Crystals, *Mol. Crys. Liq. Crys.* **63**, 111-127.

12. Migun, N. P. Experimental Method of Determining Parameters Characterizing the Microstructure of Micropolar Liquids, *J. Eng. Phys. (USSR)* **41**, 832-835.

13. Mizukami, A. Nonsteady Shear Flow of Micropolar Films, *Int. J. Eng. Sci.* **19**, 75-82.

14. Niefer, R. A., and Kaloni, P. N. Motion of a Rigid Sphere in a Shear Field in a Micropolar Fluid, *Int. J. Eng. Sci.* **19**, 959-966.

15. Nikolayevskiy, V. N. Derivation of Continuum Models of Media with Microstructure by the Method of Averaging, *Fluid Mech. Sov. Res.* **10**, 108-126.

16. Petrov, M., and Durand, G. On the Origin of Electrohydrodynamic

Instabilities in Smectic C Liquid Crystals, *J. Phys. (Paris) Lett.* **42**, 519-522.

17. Rao, S. K. L., and Iyengar, T. K. V. The Rectilinear Oscillations of a Spheroid in a Micropolar Fluid, *Int. J. Eng. Sci.* **19**, 161-188.

18. Rao, S. K. L., and Iyengar, T. K. V. Slow Stationary Flow of Incompressible Micropolar Fluid Past a Spheroid, *Int. J. Eng. Sci.* **19**, 189-220.

19. Sawada T., Tanahashi, T., and Ando, T. Oscillatory Flow of a Fluid with Couple Stress, *Keio Sci. Technol. Rep.* **34**, 65-82.

20. Sawada, T., and Tanahashi, T. Fundamental Steady Flow of Polar Fluids, *Bull. JSME* **24**, 1778-1786.

21. Shukla, P. L. Polar Flow Induced by Prescribed Injection in an Infinite Channel, *Acta Mech.* **40**, 63-73.

22. Singh, C., and Sinha, P. Dynamic Loading of Micropolar Fluid Lubricated Short Journal Bearings, *J. Mech. Eng. Sci.* **23**, 37-44.

23. Sinha, P., and Singh, C. The Effect of Additives in the Lubricant of a Composite Bearing with an Inclined Stepped Surface, *Wear* **66**, 17-26.

24. Sinha, P., Singh, C., and Prasad, K. R. Effect of Viscosity Variation Due to Lubricant Additives in Journal Bearings, *Wear* **66**, 175-188.

25. Tandon, P. N., and Rakesh, L. Effects of Cartilage Roughness on the Lubrication of Human Joints, *Wear* **70**, 29-36.

26. Yadav, J. S., and Kapur, V. K. Inertial and Fluid Suspension Effects in a Thrust Bearing, *Wear* **66**, 9-16.

27. Zaheeruddin, Kh. The Dynamic Behavior of Squeeze Films in One-Dimensional Porous Journal Bearings Lubricated by a Micropolar Fluid, *Wear* **71**, 139-152.

28. Wissbrun, K. F. Rheology of Rod-Like Polymers in the Liquid Crystalline State, *J. Rheol.* **25**, 619-662.

1982

1. Ahmadi, G. Continuum Theory of Smectic A Liquid Crystals, *J. Rheol.* **26**, 535-556.

2. Barratt, P. J., and Zuniga, I. Theoretical Investigation of the Pieranski-Guyon Instability in Couette Flow of Nematic Liquid Crystals, *J. Non-Newtonian Fluid Mech.* **11**, 23-36.

3. Carlsson, T. Possibility of the Existence of a Positive Leslie Viscosity α_2. Proposed Flow Behavior of Disk Like Nematic Liquid Crystals, *Mol. Cryst. Liq. Cryst.* **89**, 57-66.

4. Dwivedi, A. P., Pal, T. S., and Rakesh, L. Micropolar Fluid Model for Blood Flow Through a Small Tapered Tube, *Indian J. Technol.* **20**, 295-299.

5. Nigam, K. M., Jaggi, S., and Manohar, K. Micropolar Fluid Lubrication Between Two Parallel Plates with Reference to Human Joints, *Int. J. Mech. Sci.* **24**, 661-671.

6. Sastry, V. U. K., and Rao, V. R. Numerical Solution of Micropolar Fluid Flow in a Channel with Porous Walls, *Int. J. Eng. Sci.* **20**, 631-642.

7. Singh, C. Lubrication of a Roller on a Plane in Combined Rolling, Sliding and Normal Motion with Additives, *Int. J. Mech. Sci.* **24**, 495-508.

8. Singh, C., and Sinha, P. The Three-Dimensional Reynolds Equation for Micropolar Fluid-Lubricated Bearings, *Wear* **76**, 199-209.

9. Sinha, P., and Singh, C. Micropolar Squeeze Films Between Rough Rectangular Plates, *Appl. Sci. Res.* **39**, 167-179.

10. Sinha, P., and Singh, C. Micropolar Squeeze Films in Porous Hemispherical Bearings, *Int. J. Mech. Sci.* **24**, 509-518.

11. Sinha, P., and Singh, C. Lubrication of Rough Surfaces — A Microcontinuum Analysis, *Int. J. Mech. Sci.* **24**, 619-633.

12. Sinha, P., Singh, C., and Prasad, K. R. Microcontinuum Analysis of the Self Propulsion of the Spermatozoa in Cervical Canal, *Int. J. Eng. Sci.* **20**, 1037-1048.

13. Sinha, P., Singh, C., and Prasad, K. R. Lubrication of Human Joints — A Microcontinuum Approach, *Wear* **80**, 159-181.

14. Somov, A. N., and Churaev, N. V. Boundary Layer of a Nematic Liquid Crystal in Fine Capillaries, *Colloid J. USSR* **44**, 556-557.

15. Volkov, V. S. Slow Motion of a Sphere Through an Anisotropic, Viscoelastic Fluid, *J. Appl. Math. Mech. (USSR)* **46**, 185-189.

1983

1. Bhattacharyya, S. P., and Jena, S. K. On the Stability of a Hot Layer of Micropolar Fluid, *Int. J. Eng. Sci.* **21**, 1019-1024.

2. Chandra, P., and Agarwal, R. P. Dispersion in Simple Microfluid Flows, *Int. J. Eng. Sci.* **21**, 431-442.

3. Devanathan, R., and Parvathamma, S. Flow of Micropolar Fluid Through a Tube with Stenosis, *Med. Biol. Eng. Comput.* **21**, 438-445.

4. Dey, J., and Nath, G. Incompressible Micropolar Fluid Flow Over a Semi-Infinite Plate, *Int. J. Eng. Sci.* **21**, 967-972.

5. Gorla, R. S. R. Micropolar Boundary Layer Flow at a Stagnation Point on a Moving Wall, *Int. J. Eng. Sci.* **21**, 25-33.

6. Gorla, R. S. R., Pender, R., and Eppich, J. Heat Transfer in Boundary Layer Flow Over a Flat Plate, *Int. J. Eng. Sci.* **21**, 791-798.

7. Hiremath, P. S. Flow of Micropolar Fluid Through a Channel with Injection, *Acta Mech.* **46**, 271-279.

8. Kaloni, P. N. A Note on the Drag Force on a Spherical Particle in a Quadratic Flow of a Micropolar Fluid, *Lett. Appl. Eng. Sci.* **21**, 1001-1008.

9. Kolpashchikov, V. L., Migun, N. P., and Prokhorenko, P. P. Experimental Determination of Material Micropolar Fluid Constants, *Int. J. Eng. Sci.* **21**, 405-411.

10. Olmstead, W. E., and Majumdar, S. R. Fundamental Oseen Solution for the 2-Dimensional Flow of a Micropolar Fluid, *Int. J. Eng. Sci.* **21**, 423-430.

11. Ramkissoon, H., and O'Neill, M. E. Stokes Force in Continuum Fluid Mechanics, *Acta Mech.* **46**, 255-263.

12. Rao, K. M., and Rao, B. K. Longitudinal Wave Propagation in a Micropolar Wave Guide of Circular Cross Section, *Lett. Appl. Eng. Sci.* **21**, 1137-1144.

13. Rao, S. K. L., and Iyengar, T. K. V. Rotary Oscillations of a Spheroid in an Incompressible Micropolar Fluid, *Int. J. Eng. Sci.* **21**, 973-987.

14. Rao, S. K. L., Raju, K. V., and Iyengar, T. K. V. Uniqueness of Compressible Micropolar Fluid Flows, *Int. J. Eng. Sci.* **21**, 143-153.

15. Sastry, V. U. K., and Rao, V. R. Numerical Solution of Thermal Instability of a Rotating Micropolar Fluid Layer, *Int. J. Eng. Sci.* **21**, 449-461.

16. Sinha, P., and Singh, C. Theoretical Effects of Rigid Particle Additives in Noncyclic Squeeze Films, *J. Lubr. Technol.* **105**, 105-112.

17. Sinha, P., and Singh, C. Effect of Lubricant Additives on Finite Step Bearings, *Wear* **85**, 1-9.

18. Sinha, P., and Singh, C. Microcontinuum Analysis of Squeeze Films Between Rough Surfaces, *Wear* **89**, 69-82.

19. Sinha, P., Singh, C., and Prasad, K. R. Viscosity Variation Considering Cavitation in a Journal Bearing Lubricant Containing Additives, *Wear* **86**, 43-56.

20. Soundalgekar, V. M., and Takhar, H. S. Flow of Micropolar Fluid Past a Continuously Moving Plate, *Int. J. Eng. Sci.* **21**, 961-965.

21. Yousif, A. E., and Ibrahim, T. M. Lubrication of Infinitely Long Thrust Bearings with Micropolar Fluids, *Wear* **89**, 137-150.

22. Zaheeruddin, Kh., and Ayyubi, S. K. Theoretical Study of the Effects of Solid Particles in the Lubricant of Externally Pressurized Porous Thrust Bearings, *Wear* **89**, 237-244.

Subject Index

Experiments in Fluids

Experimental Methods and
Their Applications to Fluid Flow

ISSN 0723-4864 Title No. 348

Experiments in Fluids publishes research papers and technical notes which describe either:

- The development of new measuring techniques or the extension and improvement of existing methods for the measurement of flow properties necessary for the better understanding of fluid flows and their application in science and engineering.
- The application of experimental methods to the solution of problems of fluid flow.

The contributions encompass a wide range of applications including aerodynamics, hydrodynamics, basic fluid dynamics, convective heat transfer, combustion, chemical, biological, and geophysical flows, and turbomachinery. Those which report advances to the analyses of flow problems will be considered provided they contain a substantial experimental content.

Subscription information and/or **sample** copies are available from your bookseller or directly from Springer-Verlag, Journal Promotion Dept., P.O.Box 105280, D-6900 Heidelberg, FRG

Springer-Verlag
Berlin
Heidelberg
New York
Tokyo

MIX
Papier aus verantwortungsvollen Quellen
Paper from responsible sources
FSC® C105338
FSC
www.fsc.org

If you have any concerns about our products,
you can contact us on
ProductSafety@springernature.com

In case Publisher is established outside the EU,
the EU authorized representative is:
Springer Nature Customer Service Center GmbH
Europaplatz 3, 69115 Heidelberg, Germany

Printed by Libri Plureos GmbH
in Hamburg, Germany